The Economics of Soilless Production

THE ECONOMICS OF SOILLESS PRODUCTION

Nekesah T. Wafullah

Delve
PUBLISHING

www.delvepublishing.com

The Economics of Soilless Production

Nekesah T. Wafullah

Delve Publishing
224 Shoreacres Road
Burlington, ON L7L 2H2
Canada
www.delvepublishing.com
Email: orders@arclereducation.com

ISBN: 978-1-77469-535-7 (Hardcover)

ABOUT THE AUTHOR

Nekesah T. Wafullah is a skilled agriculture expert with extensive knowledge in agricultural energy value addition products, agricultural business management services, project management, various forms of fertilizer, their production, sales, marketing aspects and application regimes; cross border fertilizer trade policies; youth and women empowerment and volunteerism. She is adept at project planning and management as well as creating simple solutions to complex problems. She has experience within agricultural markets in Kenya, Zambia, Malawi, Tanzania, Rwanda, Democratic Republic of Congo (DRC)- Bukavu and Lubumbashi and Uganda. She mentors high school and college students and advocates for better performance in Agricultural science. During her free time, she loves editing books, watching movies, cooking, baking, networking, reading, and dancing. Nekesah holds an M Sc. in Agricultural and Applied Economics degree from the University of Nairobi with a major in International Trade and Policy.

TABLE OF CONTENTS

LIST OF FIGURES

LIST OF TABLES

LIST OF ABBREVIATIONS

AD	Anaerobic Digestion
AEC	Anion Exchange Capacity
AFP	Air-Filled Porosity
BD	Bulk Density
BNF	British National Formulary
CA	Contact Angle
CEA	Controlled Environment Agriculture
CEC	Cation Exchange Capacity
CGW	Composted Green Waste
DFT	Deep Flow Technique
DWC	Deep Water Culture
EAW	Easily Available Water
EC	Electrical Conductivity
ET	Evapotranspiration
EU	European Union
FAO	The Food and Agriculture Organization of The United Nations
FTE	Fritted Trace Elements
GHG	Greenhouse Gas
HID	High-Intensity Discharge
HPS	High-Pressure Sodium
HVAC	Heating, Ventilation, And Air Conditioning
LCA	Life Cycle Analysis
LED	Light-Emitting Diode
NFT	Nutrient Film Technique
NS	Nutrient Solution
NVZs	Nitrate Vulnerable Zones
OM	Organic Matter
OUR	Oxygen Uptake Rate
pH	Potential of Hydrogen
PSD	Particle-Size Distribution

PVC	Polyvinyl Chloride
PW	Pore-Water
QTK	Quick Test Kits
RC	Retention Curve
RSMS	Rhizon-Soil-Moisture-Sampler
SAIs	Specifically Adsorbed Ions
SEM	Scanning Electron Microscopy
SWR	Soil-Water Repellency
TDR	Time-Domain Reflectometry
TOV	Tomato-On-Vine
TP	Total Porosity
TVCs	Total Viable Count
UC	University of California
US	United States
USD	United States Dollar
UV	Ultra-Violet
WRAP	The Waste and Resources Action Program
WRC	Water Retention Curve
WU	Water Uptake

ACKNOWLEDGEMENT

The book is the product of great effort and time spent. The completion of this book could not have been possible without the participation and assistance of so many people whose names may not all be enumerated. Their contributions are sincerely appreciated and gratefully acknowledged. However, I would like to express our deep appreciation and indebtedness, particularly to the following: Arcler Education, Inc. and Charles Kuria for their endless support, kind and understanding spirit during this undertaking. To all relatives, friends and others who in one way or another shared their support, either morally, financially or physically, I would like to personally thank you.

PREFACE

Plants grow in soil, as we all know, but what exactly is soil? As growing plants in a soilless medium is becoming increasingly popular around the world because of a lack of suitable soils, disease contamination from repeated use, and the desire to apply the best conditions for plant growth. There is a lot to discuss when it comes to soilless agriculture. In soilless culture, we just provide these same nutrients that have been pre-mixed into a water reservoir, coupled with high amounts of oxygen, to the plant roots efficiently in soilless agriculture, such as hydroponics. Although soilless cultivation is typically thought of as a recent technique, cultivating plants above ground in containers has been attempted numerous times throughout history. Different soilless agriculture advancement has been enhanced by different forces. These forces include the quickening of society's scientific, economic, and technological advancements.

Population growth around the world and rising living standards in many nations have increased the demand for high-quality, out-of-season vegetables as well as high-value meals and ornamentals. With the rise in living standards, demand for floricultural products, such as cut flowers, potted plants, and bedding plants, has also increased significantly. Growers had to enhance their production intensity to maintain their competitiveness because protected farming has higher production costs than open field growing. This was accomplished through a variety of strategies, chief among which is the sharp rise in soilless production in relation to total agricultural crop production. The spread of soil-borne infections in intensively farmed greenhouses was a major factor in the decision to abandon the use of soil. The transition from soil to substrate growing systems was accelerated by the restriction on the use of methyl bromide, which is thought to be an ozone layer lowering agent.

Different substrates, such stone wool, polyurethane, perlite, scoria (tuff), etc., have taken the place of soil because they are essentially free of pests and diseases thanks to the production process. Additionally, these materials can be disinfested during reuse from crop to crop to eradicate any germs. The ongoing trend toward soilless farming is also fueled by the fact that in soilless systems, it is feasible to exert more direct control over a number of important variables, which significantly enhances plant performance. Most substrates have better physical and hydraulic properties than soils. After irrigation, a plant cultivated on soil has comparatively high-water availability. Despite the many obstacles, it is evident that this is an area of agriculture that is expanding quickly. Common crops grown under soilless culture include leafy vegetables, fruit vegetables and fruits like strawberries. This book looks at the details of most components used currently in soilless culture, with the aim of promoting awareness among the audience and guiding readers on the best soilless strategies that pay off.

CHAPTER 1

INTRODUCTION TO SOILLESS PRODUCTION

CONTENTS

OVERVIEW

Plants grow in soil, as we all know, but what exactly is soil? It provides the standard macro and micronutrients required for plant growth, making it our most abundant growing medium. These nutrients are carried to the roots of the plant by water as it passes through the soil. We just provide these same nutrients that have been pre-mixed into a water reservoir, coupled with high amounts of oxygen, to the plant roots efficiently in soilless agriculture, such as hydroponics. There is a lot to discuss when it comes to soilless agriculture, thus this section will primarily concentrate on the common soilless cultivation techniques you can utilize, their benefits, and a brief history of them.

HISTORY AND CONCEPT OF SOILLESS PRODUCTION

Although soilless cultivation is typically thought of as a recent technique, cultivating plants above ground in containers has been attempted numerous times throughout history. It was done by the Egyptians over 4,000 years ago. The Deir el Bahari temple's wall murals revealed what appears to be the earliest instance of container-grown plants in history. When the local soils were unsuitable for the particular plant, they were employed to transport mature trees from their original nations to the king's palace and then to be cultivated there. The sort of growing medium used to fill the containers is unknown, however, given that they were proved to be transported across long distances by porters, it's probable that the components utilized were lighter than pure soil. Beginning in the 17th century, plants were transported, primarily from the Middle East and the Far East to Europe to be grown in orangeries, in order to give exotic fruits and vegetables and items of aesthetic value to affluent people. Although soil was mostly used to fill these containers, an orangery can be regarded as the first instance of a greenhouse-container-grown system because it was "a covered area, notably a greenhouse, utilized for the growing of orange trees in cool regions." Today, orangeries can still be seen all over Europe. A beautiful illustration from Moscow, Russia. Citrus trees cultivated in containers were winter-protected at Pill-Nitz Palace, close to Dresden, Germany. Large doors on the east side made it possible to transfer trees inside and outside, allowing them to be grown outside in the summer and brought inside in the winter. The south side had large floor-to-ceiling windows that let light in. As implied by the name, various citrus species were the first plants to be grown in orangeries. The development of the production of plants in containers required the completion of two significant phases. One was the discovery of the nutritional needs of plants, which was made possible by French and German scientists in the 19th century and later developed by English and American scientists in the first part of the 20th century. Although it is feasible to grow plants in silica sand using nutrient solutions, British scientists in 1946 nevertheless maintained that plants grown in similarly treated soil produced more yield and biomass (Woodman and Johnson, 1946a, b). In order to show the superiority of soilless media in terms of yield, researchers did not develop complete nutritional solutions until the 1970s. They then combined its use with suitable rooting media and explored how to maximize the quantities of nutrients, water, and oxygen (Cooper, 1975; Verwer, 1977).

Figure 1: The orangeries at Pill-Nitz Palace.

Source: Wikimedia Commons

The second crucial step was realizing that while soil-grown plants were nearly difficult to rid themselves of disease organisms that needed to be controlled by disinfestation, eliminating them from container-grown produce was doable. The description of a production system that supplied a handbook for the use of substrates in conjunction with disease control for the production of container-grown plants in outdoor nursery production was a major publication in the United States. The breakthrough in container nursery production during the 1950s and 1960s was known as "The U.C. System for Producing Healthy Container-grown Plants through the use of Clean Soil, Clean Stock, and Sanitation" (Baker, 1957). Because growers who used the system had a significant financial advantage over rivals who did not, it was widely adopted. This guidebook specified numerous combinations of sand and organic material, such as peat, bark, or sawdust, in varying amounts. These were dubbed "UC mixes." It should be noted that these mixtures were referred to as "soil" or "soil mixes" in this text, partly because most container media up until that point were composed of a mixture of soil and different elements. This book does not follow that practice; instead, we define "soil" as a specific mixture of sand, silt, clay, and organic stuff present in the earth.

As a result, when we refer to soilless substrates in this book, we often mean mineral components (such sand or clay) that are also present in soil but do not actually include any soil. For many years, "compost" was also used interchangeably with "soil mix," mostly in Europe and the United Kingdom but also in the United States (Robinson and Lamb, 1975). (Boodley and Sheldrake, 1973). This phrase encompassed what is now often referred to as substrate or growing medium and, in most instances, implies the utilization of a mixture of various components, at least one of which is of organic origin. In this book, the terms "growing medium" and "substrate" are used synonymously, whereas "compost" refers to organic matter that has undergone a composting process in order to stabilize it and eradicate plant and zoonotic pathogens. The aforementioned scientific breakthroughs disproved the theory that growth media may be put just by randomly mixing some soil and other components to form "potting soil." This idea has historically been backed by the fact that creating the optimal growing media was mostly an iterative process. Physical Characteristics of Soilless Media and Chemical Characteristics of Soilless Media, which are obtained through the combination of particular components (for example, UC mix) or through industrial manufacture (for example, stone-wool slabs), respectively, give us a fairly complete picture of the significant physical and chemical characteristics that are achieved. There are several local and regional applications of these ideas all around the world. These are typically motivated by horticultural and monetary factors.

In the end, the primary factor influencing the composition of a specific substrate mix that is used in a soilless production scenario is financial concerns. The availability of supplies, the cost of processing, the cost of transportation, and the cost of producing plants or crops, as well as the costs involved with their transportation and marketing, are all examples of how the financial aspect expresses itself. In some circumstances, disposing of spent substrates is a crucial factor to take into account while analyzing the environmental and financial effects. For instance, because mineral wool (stone and glass wool) is not a natural resource that can be recycled, its safe disposal is one of the main issues with its usage in horticulture. Diverse techniques for recycling stone-wool have been devised, however, they all place some financial burden on the end-user. The depletion of natural resources is another environmental aspect that could restrict the usage of different materials as growth substrates. As a result, there is a growing movement throughout most of Europe to reduce the mining of peat, one of the most common substrates in temperate climates (Alexander et al., 2017).

Growers discover this material to be less expensive in nations where peat is easily accessible, possibly even locally harvested, than in nations where it must be imported from far-off places. Growers must decide whether to employ a "tried and true" component (like peat) or a replacement (like coconut coir) in a recipe that may have proven successful over time when raw material prices change. The financial condition can prompt contemplation of a change in some years. Since the characteristics of each substrate and mixture vary, switching out a specific component, like peat, may increase costs or produce crops of lower quality, which may be worth less on the market. This is especially true if the grower is switching to a material with which they are less familiar. As a result, the problem facing farmers worldwide is to put up mixtures that will perform as intended at the lowest feasible overall cost. As a result, while making an effort to follow a particular set of rules, the substrates utilized globally have very different compositions. These rules, which relate to physical and chemical elements of solids, liquids, and gases in the plant's root zone are highly intricate. The production of ornamentals and vegetables in greenhouses, as well as outdoor container nurseries, are now the two sectors of industry where soilless production is most prevalent. Virtually all containerized plants are grown in urban gardening without any field soil.

THE CONCEPT OF HYDROPONICS

Water culture without the use of any solid substrates has also been used to grow plants without soil. The term "hydroponics" is widely used to describe this sort of soilless agriculture. Gericke (1937) originally used this word to refer to water culture without the need for a substrate, but other people now use it to indicate different things. Many people use the phrase to describe systems that do contain a substrate of some kind to anchor or stabilize the plant and to offer an inert matrix to store water. However, technically speaking, hydroponics refers to the process of growing plants in nutrient solutions. We also include in this idea substrate-based systems where the substrate does not provide nutrients or ionic adsorption or exchange, such as those that use the Nutrient Film Technique (NFT), Deep Flow Technique (DFT), and aero-hydroponics. So, we define hydroponic production systems as those that use inert substrates like stone wool or gravel. In this book, we will typically use the term "hydroponics" in conjunction with qualifying terms to distinguish between "substrate-based hydroponics" and "liquid-culture hydroponics"; the latter involves a significant amount of substrate that is inert and has a little ion-exchange capacity, whereas the former has no (or almost no) substrate. Hydroponics was first primarily utilized by scientists as a research tool to examine certain aspects of plant nutrition and root activity. The development of numerous types of substrates, in addition to advancements in the production of totally soluble fertilizers, automation, and the manufacture of polymers, combined with scientific advances to make soilless farming a feasible commercial endeavor. In greenhouses nowadays, a variety of soilless systems are available for the cultivation of vegetables and ornamentals. As a result, there are many different growth systems.

IMPORTANCE OF SOILLESS PRODUCTION IN AGRICULTURE

Since the forces that sparked these changes are still in existence, the world of agriculture has undergone significant transformation over the previous few decades and will continue to do so. These forces include the quickening of society's scientific, economic, and technological advancements. Population growth around the world and rising living standards in many nations have increased the demand for high-quality, out-of-season vegetables as well as high-value meals and ornamentals. With the rise in living standards, demand for floricultural products, such as cut flowers, potted plants, and bedding plants, has also increased significantly. These developments led to an increase in the usage of a wide range of covered agricultural methods, from simple plastic or screen covers to fully climate-controlled greenhouses. This manufacture was initially done wholly in the ground, where the soil had been altered to promote good drainage. Growers had to enhance their production intensity to maintain their competitiveness because protected farming has higher production costs than open field growing. This was accomplished through a variety of strategies, chief among which is the sharp rise in soilless production in relation to total agricultural crop production. The spread of soil-borne infections in intensively farmed greenhouses was a major factor in the decision to abandon the use of soil. The transition from soil to substrate growing systems was accelerated by the restriction on the use of methyl bromide, which is thought to be an ozone layer lowering agent. Different substrates, such as stone wool, polyurethane, perlite, scoria (tuff), etc., have taken the place of soil because they are essentially free of pests and diseases thanks to the production process. Additionally, these materials can be disinfested during reuse from crop to crop to eradicate any germs. The ongoing trend toward soilless farming is also fueled by the fact that in soilless systems, it is feasible to exert more direct control over a number of important variables, which significantly enhances plant performance. Most substrates have better physical and hydraulic properties than soils. After irrigation, a plant cultivated on soil has comparatively high-water availability. The macropores in the soil are now filled with water, followed by a somewhat sluggish drainage process that allows air to enter the soil macropores. The oxygen that soil microflora and plant roots use up is replaced at a rate that may be slower than what plants need. The porosity of the soil is sufficient that air oxygen diffuses into the root zone as water is drained or lost to evapotranspiration. The plant must expend a lot of energy to absorb enough water to make up for transpiration losses caused by atmospheric demand because some water is simultaneously trapped by steadily increasing soil matric pressures. Contrarily, the majority of substrates permit simultaneous optimization of oxygen and water availability. In contrast to soil, substrates have much less matric forces holding the water. Therefore, plants growing in porous media at or close to container capacity use less energy to get water from the soil. At the same time, a sizeable portion of the macropores is filled with air, ensuring that the oxygen diffusion rate is high enough to prevent the risk of oxygen deficit in plants, which is something that happens to plants growing in soil that is close to the field capacity. Another factor is that soilless agriculture allows for greater management and control of nutrient supply to plant roots than most arable soils do.

Growing plants in a soilless medium is becoming increasingly popular around the world because of a lack of suitable soils, disease contamination from repeated use, and the desire to apply the best conditions for plant growth. The majority of these plants are raised in greenhouses, typically in close to ideal production environments. The fact that the root volume is constrained in containerized culture as opposed to soil-based culture is a disadvantage of soilless cultivation over the latter. This constrained root volume has a number of significant implications, most notably a constrained supply of nutrients and water. Since there are more roots per unit volume of media, the constrained root volume also enhances root-to-root rivalry. In terms of pollution control, container production systems are preferable to in-ground production systems because they allow for the reduction or elimination of nutrient ions and pesticide residue discharge, protecting freshwater lakes. In addition, container production often has far improved water and nutrient use efficiencies, which has obvious cost advantages. Environmental pollution reduction is receiving increasing attention across all developed nations, and in those where this type of production accounts for a significant portion of agricultural productivity, regulations are being created to compel recirculation in order to reduce or completely eliminate run-off from nurseries and greenhouses. Many studies are being conducted right now on the benefits and limitations of closed and semi-closed systems. In many instances, professionals approach irrigation as distinct from fertilization, which is then treated separately from the planning and creation of the substrate used to grow the plants. This chapter discusses the root zone as a dynamic system and demonstrates how such a system is constructed and controlled in order to maximize crop production while also taking into account socially imposed constraints (run-off elimination, labor savings, etc.). The new trend of "organic hydroponics," which appears to be gaining popularity in some regions of the world, is another topic covered in this chapter. The production of sufficient amounts of food at reasonable prices in less developed nations is one of the major difficulties facing global horticulture in the future. The issues brought on by the absence of fertile soils and knowledge may be partially solved by simple, affordable soilless production technologies. This development can be encouraged by the fact that a relatively small farmed area can feed a big population. As a result, specialists should be encouraged to create cheaper, more durable alternatives to the existing high-tech, expensive equipment and procedures that are used in distant places. In this regard, one of the most significant benefits of soilless farming is to be mentioned: in the majority of developing nations, water is both scarce and of poor quality.

The FAO's hunger map and aridity index map make it evident that a significant portion of the population experiences hunger in many parts of the world, primarily owing to water scarcity in places like sub-Saharan Africa, Namibia, Mongolia, etc. Since soilless plant cultivation (and particularly in recirculated systems) uses less water than soil-grown plant production, more food can be produced with such methods. Additionally, compared to plants grown in soil, plants growing in such systems may tolerate higher saline levels better. This is due to the relationship between the roots' capacity to keep out harmful ions like Na1 and their ability to resist high osmotic pressure. In this regard, it is interesting to note that soilless farming is widely used in dry regions like the majority of Australia, several sections of South Africa, Saudi Arabia, and the Southern part of Israel. There is no issue with hunger in any of these nations. The growing concern in sustainability and circularity is another aspect. One apparent aspect is to use human or animal urine and manure in some way to produce nutrient solutions for plants, either to produce fertilizer systems or goods directly (Wald, 2017). Of course, this is a throwback to older times when farmers would spread animal feces on fields to serve as plant fertilizer. This is no longer done in a lot of the developed world since it contaminates the water supply. But in many parts of the poor world, this continues to be the primary fertilizer used in agricultural crops. It is worthwhile to investigate how such circularity might be reintroduced to the existing, extremely modern manufacturing processes that completely prohibit this. The fact that these fertilizers are not ionic makes it difficult for the current sensors to work. A system called aquaponics, which combines hydroponics and aquaculture, is one where this kind of circularity is imposed directly. According to the notion, aquatic organisms can be fed unmarketable plant material as food and the waste from aquaculture can be utilized by plants. The main issue is whether the two full cycles can be run in a way that promotes healthy growth in both aquatic and terrestrial organisms. Unfortunately, there are still several obstacles in the way of getting these two cycles to interact. While some difficulties (such as the management of ammonium) are reasonably well understood, other difficulties still need creative new solutions.

The utilization of plant-growing systems that do not expose the plants to direct sunlight but rather utilize electricity to create photosynthetically active radiation is another new development in controlled environment agriculture. With the development of light-emitting diode lamp technologies, plant growth lighting can efficiently provide irradiance for the plants that is fully utilizable by the plants by converting energy into photons of precise wavelengths. The shift toward more affordable lighting will make more of these systems practical. Since it can only be done if every variable is optimized, hydroponics is always a component of such systems. Another trend in indoor controlled environment agriculture is the vertical stacking of production systems, which is made possible by the use of fully automated hydroponic systems in production systems that use LED-generated lamps as light sources. As a means of integrating contemporary production systems into cities, vertical farms have grown to be quite intriguing. Despite the many obstacles, it is evident that this is an area of agriculture that is expanding quickly. Condensate from the air conditioning system, a novel sort of water not previously available in the field or greenhouse plant cultivation, is also present in such systems. It is conceivable to attain water-use efficiencies in these systems with complete recirculation of all water molecules that are unimaginable in field agriculture. The potential to grow plants in harsh deserts and polar environments (where few plants now grow) is thus presented, and all of the knowledge in this book will be very helpful in attaining success. Although significant research has been done, the science of plant production in soilless systems is still in its infancy, and many problems remain unresolved. In order to maximize the use of the currently available substrates and to suggest desirable qualities for future substrates and growing techniques to be developed by the next generation of researchers, one of the goals of this book is to concentrate on the key issues of the biological, physical, and chemical environment of the rhizosphere.

ADVANTAGES OF SOILLESS AGRICULTURE

Faster growth: A better feeding strategy will result in more grow cycles per year, a higher yield, and more food overall.

Extreme decrease in water and nutrient use: Hydroponics typically utilizes at least 90% less water than soil-based techniques since it closes the loop and circulates the water through the system. Only a small amount of the water and nutrients used in conventional outdoor soil farms actually reach the roots; the remainder is lost to the local water supply.

Controlled Environment Agriculture (CEA): Using CEA, a farm can optimize its growth environment by controlling the temperature, humidity, ventilation, and light. Anywhere in the world, growers utilizing greenhouses or indoor vertical farms can generate ideal growth conditions every day of the year. This is how soilless agriculture is frequently practiced. Additionally, a farm's capacity to forecast crop timing, produce high-quality plants, and uphold strict food safety guidelines is greatly enhanced by growing in a controlled environment. Pesticides may be avoided, and there aren't any birds soaring overhead or creatures burrowing in the area. To avoid or handle any pest issues, many CEA farms use integrated pest control (predatory bugs). Although the farm's ability to produce will not be affected by the weather, it will use more energy than conventional farms (due to lighting and HVAC).

Location: The farm can and should be much closer to the final consumer, reducing transportation emissions and enhancing product freshness. This type of growing can be done anyplace because it is not reliant on fertile soil.

Use significantly less space: The yield per square foot significantly increases with stacked or vertical systems. This is essential if we want to be able to feed an expanding population without having to convert more natural areas into farming, such as wetlands, forests, or other significant ecosystems.

DISADVANTAGES OF SOILLESS AGRICULTURE

Drying Out: The majority of soilless potting mixes are made out of peat moss, coconut coir, sand, compost, perlite, and vermiculite, and none of them can retain water for very long. Their major drawback is also a result of their superior drainage and air room for potted plants: In the scorching summer weather, a little potted plant in a soilless potting mix dries out in less than a day, necessitating continuous watering. These plants need to be submerged in a water basin to become rehydrated after becoming fully dry and hydrophobic.

One-time equipment investment is large: The price of fertilizer is high, the amount of electricity used is significant, and managerial skills are necessary. For instance, certain specialized knowledge is needed for the preparation, modification, and management of nutritional liquid. Vegetable soilless culture, also known as nutrient liquid culture, is a new technique for indoor soilless growing. Its main component is to secure the plant roots in the planting basket and allow them to grow naturally into the plant nutrition solution. In place of natural soil, a liquid can supply water, nutrients, temperature, and other development conditions to the plant, allowing it to develop normally and finish its life cycle. The promotion and use of soilless agriculture techniques are now constrained by these adverse considerations. Therefore, the choice of soilless cultivation techniques and useful technologies that are straightforward, convenient, low-investment, low-cost, and good in the application will substantially enhance soilless cultivation techniques in practice, depending on the particular circumstances (Condition).

Time Consuming: While hydroponic farming may appear practical and convenient, it still takes a little bit of time. It is possible to ignore the plants growing in the soil for days or even weeks because nature and the soil have properly balanced everything. With hydroponics, this is not the case. You must be more patient with the plants growing in this phase because they nearly exclusively depend on water to survive. Regular water replacement is necessary, and you should keep an eye on the plants to make sure they're doing well.

Risks Of Water and Electricity: The two main components of hydroponic gardening are water and power. Therefore, the hydroponic system won't thrive unless you have enough water or reliable energy. You must take the necessary safety precautions when growing plants in this system to prevent any interruptions in the growth process. If you depend on electricity to run your entire hydroponic system, you should take safety precautions in case of blackouts or dimouts. It is highly important to take this precaution even before you move on with a hydroponic system because the system won't run owing to the lack of energy supply, thereby drying up the plants.

Debates Over the Organic Nature: There has been a lot of discussion and consideration over the past few years over whether hydroponic gardening is actually an approved organic farming technique. Due to the fact that hydroponically produced plants do not have access to the same microbiomes as those found in soil, several farmers have questioned whether such plants can actually be certified as organic. However, over the past ten years, people from all over the world have grown hydroponically grown tomatoes, lettuce, and other leafy vegetables. This type of farming has already been used and successfully carried out in nations like Australia, the US, and the Netherlands. It has ultimately provided nourishment for a large number of people.

TYPES OF SOILLESS SYSTEMS

There are a number of hydroponics-related approaches that enable the best possible customization of a growing operation. There is a growing method that is ideal for your objectives, whether you want to develop a sizable commercial farm or are only a hand-watering hobby grower. The best part is that you are not required to pick just one! You can combine any or all of these strategies if you have adequate room.

Deep Water Culture (DWC)

This is just a case of plants floating in water that is being circulated. There are numerous methods, however, styrofoam insulating boards are frequently used (4ft x 8ft). Your germinated plant won't fall through if you drill tapered (conical) holes in the boards the size of the growing media you choose. Until harvest, the roots are suspended in a well-oxygenated nutrition solution in a layer of 6 to 18 inches. Short-statured leafy greens and herbs are ideal because they don't need a lot of root support. DWC systems frequently store a significant amount of water, which dampens fluctuations in the solution's chemistry. Additionally, if a DWC system's pump ever breaks down, you would have plenty of time to fix it before you had any major issues, like the roots drying out.

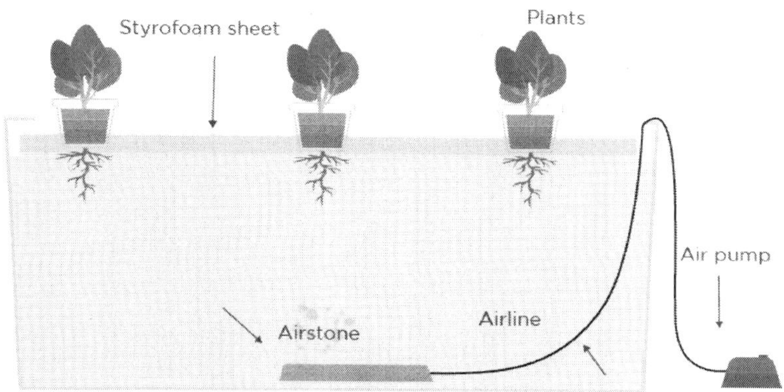

Figure 2: Deep Water Cultivation.

Source: İncemehmetoğlu, (2012).

Nutrient Film Technique (NFT)

This adaptable method involves supplying a very shallow stream of water to the roots through channels or troughs that are slightly angled for drainage purposes. It can be carried out continuously or on a timer. For ideal dissolved oxygen levels and to avoid stagnation, the solution is kept at the lowest position in a reservoir using an air stone and a submersible pump. When the roots are completely submerged, the water drains back into the reservoir. However, these systems carry significantly less water per plant and are easier to stack, clean, and customize for your grow space. NFT is great for short-statured plants like DWC.

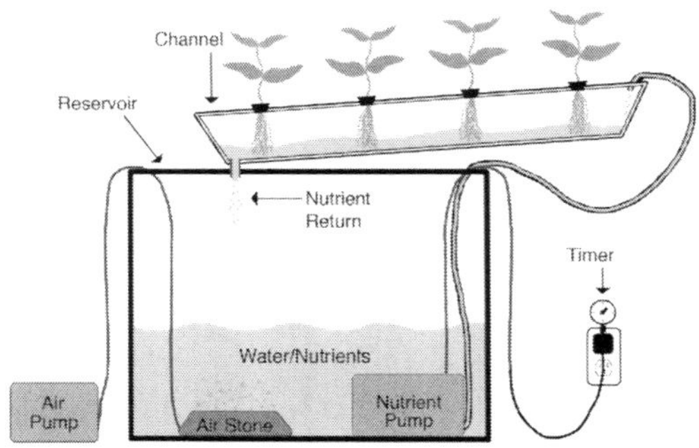

Figure 3: Nutrient Film Technique.

Source: İncemehmetoğlu, (2012).

Ebb and Flow (Flood and Drain)

A small plastic bucket filled with some expanded clay pellets or other rock media that you hand water and drain can serve as a simple ebb and flow system. It could also be as complicated as incorporating an aquaponic system with a sizable media-filled bed and flooding it with the system's liquid waste. Every time, the grow tray is briefly filled with solution, soaking the roots, and then the solution is drained back into the reservoir. Ebb and flow systems are excellent for growing just about anything, but especially fruiting crops, because of the root support and oxygen levels they can offer.

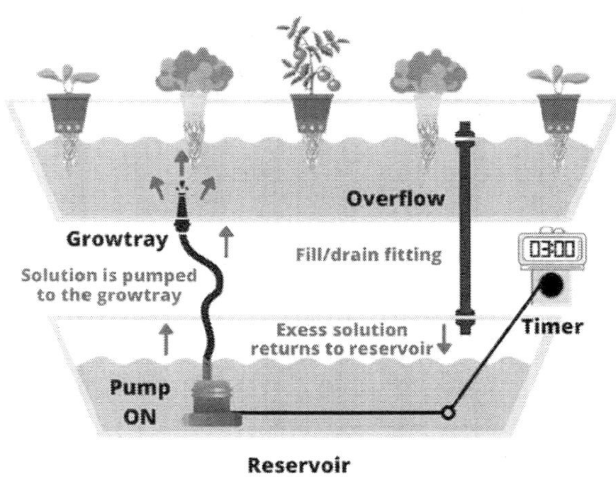

Figure 4: Ebb and Flow (Flood and Drain).

Source: İncemehmetoğlu, (2012).

Simply make sure your setup can support the weight of all that media and water, and that your containers drain completely, however it is done. Although ebb and flow systems require little maintenance and yield high yields, they are similar to NFT systems in that your plants could quickly suffer a catastrophic pump failure.

Drip System

Another popular and straightforward method is the drip system, in which a pump on a timer slowly feeds the solution to the base of each plant separately. Either the extra solution goes back into the reservoir or it's not collected. It functions well with growing media that retain a lot of water (i.e., coco coir, peat moss, or rockwool). When the system is operating well, it requires relatively little care and produces a lot of plants, but the drip lines might clog and cause plants to become dried out. For these systems, synthetic fertilizers make the most sense because organic materials quickly choke pipes.

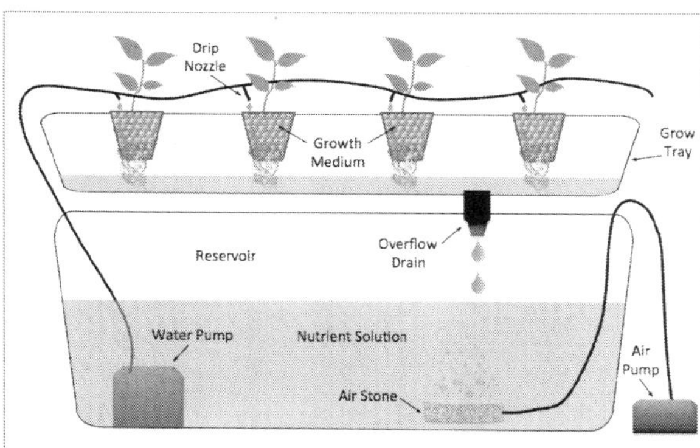

Figure 5: Drip System.

Source: İncemehmetoğlu, (2012).

Aeroponics

This novel method of spraying hydroponically grown roots hung in the air with solution. Although the fertilizer distribution and water use in these systems are incredibly exact, aeroponics may not be the best place to start when constructing your own system. Due to the emitters' extremely small openings, the clogging problems could be even severe than with a drip system. However, the roots receive plenty of oxygen, and increased oxygenation speeds up growth. A-frame aeroponic systems are widespread and use tall cones with PVC frames to create the ideal atmosphere for the root chamber.

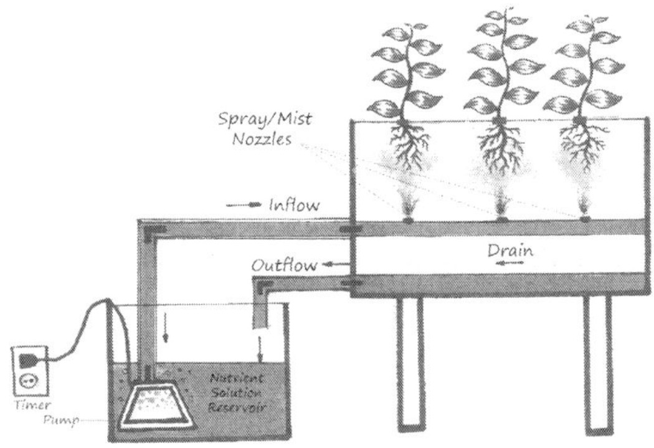

Figure 6: Aeroponics.

Source: İncemehmetoğlu, (2012).

Wicking System

Because they require little upkeep, wicking systems are the most fundamental and passive growing method and a wonderful place for newbies to start. The wicks carry the solution to a tray, and the capillary action of the plant carries it to the roots. Although it has no moving parts and performs admirably on a small scale, it is not the most effective method for using nutrients.

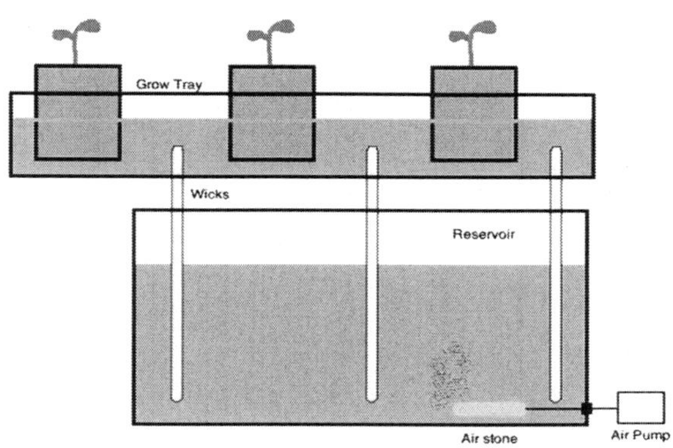

Figure 7: Wicking System.

Source: İncemehmetoğlu, (2012).

Aquaponics and Organic Hydroponics

In fact, hydroponics has been utilized and developed throughout human history for thousands of years. The Hanging Gardens of Babylon and the Floating Gardens in China both contain early examples of soilless systems. These civilizations frequently practiced aquaponics, which involves growing plants and fish in the same body of water. Actually, the fish serve as the plants' fertilizer, and the plant roots serve as the fishes' water filters. Fish emit ammonia through their gills and excrement (pee and poop!), which they produce when they are exposed to a freshwater source. Nitrifying bacteria, which are naturally occurring microorganisms, gradually start to proliferate on every underwater surface. These bacteria transform the waste from fish into nitrites and then nitrates, which are nutrients for plants. Thus, aquaponic farms actually cultivate three distinct but equally crucial things: plants, fish, and healthy bacteria. Aquaponic systems can be a very sustainable and entertaining way to generate food when the plants and fish are in the right proportions. Both organic hydroponic applications and aquaponics require the same nitrifying bacteria. Aquaponics and organic hydroponics are biologically more similar to one another than synthetic hydroponics. Since it can be a challenging and untidy process, synthetic hydroponic techniques are widely used in commercial size enterprises. One explanation for this is that keeping a healthy and active bacterial colony is crucial for organic hydroponics and aquaponics, therefore cleaning and draining much of the water is not advised and can harm the system's general health. Your water may need to "mature" for several months before the ammonia introduced by fish waste or organic fertilizers is efficiently transformed into nitrates. Since oxygen is these systems' defining component, it is crucial to have a strong flow rate and biofiltration to remove any waste that consumes oxygen to prevent anaerobic activity (absence of oxygen).

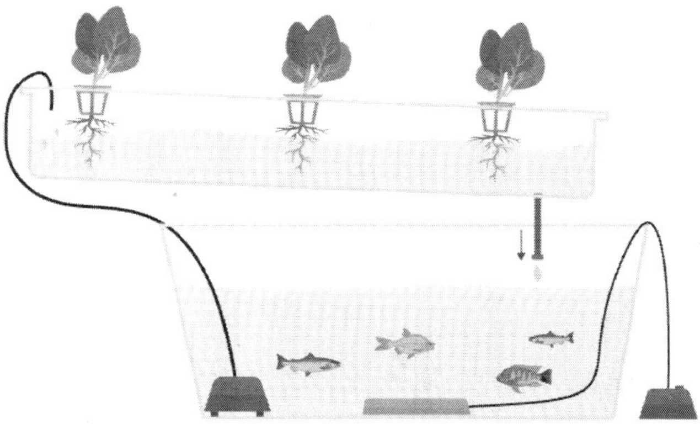

Figure 8: Aquaponics and Organic Hydroponics.

Source: Azad et al., (2016).

CHAPTER 2

SOILLESS MEDIA COMPONENTS

CONTENTS

OVERVIEW

A growing mix's main function is to contain water and nutrients for the plant to use while it is growing while also supporting the plant. The physical and chemical characteristics of the growth mix will be influenced by the ingredients chosen to manufacture it. We'll examine the most frequently used ingredients added to growing media and how they affect the characteristics of the resulting medium. We'll examine organic, artificial, and inorganic elements. As a percentage of the mix, these ingredients are blended in different proportions. Due to the wide range of origins, physical and chemical features, and the regular introduction of new products, it is impossible to cover all substrate types that are currently on the market. Therefore, only the media that are typical and commonly used are described here.

INORGANIC AND SYNTHETIC ORGANIC COMPONENTS

Introduction

Field soils and soilless substrates share the same fundamental principles, but have very distinct physical and hydraulic qualities. Additionally, there are significant distinctions between spatially constrained soilless production systems and unconfined field soils in terms of dynamic processes for water, air, and nutrient delivery as well as root growth and development. While water drainage and nutrient distribution are constrained in growing containers, these processes are unimpeded in agricultural soils where water drains and redistributes to considerably deeper layers. Due to the drastically altered infiltration and redistribution dynamics that result, soilless systems need to be managed more intensively in order to maintain a stress-free rhizosphere environment for the best plant growth. For intensive soilless production systems, factors like substrate chemistry and plant nutrition are just as crucial as physical factors. Although the fundamentals of mineral plant nutrition, which have been thoroughly reviewed by a number of authors (Kirkby, 2001; Epstein and Bloom, 2005), are generally the same as those of soil-based production, some factors, primarily those relating to volume constraints, call for special attention. Reduced buffer capacity for provided nutrient solutions and associated limited nutrient supply due to volume limits must be taken into account for proper fertilization management (capacity factor). The components of horticultural soilless substrates and potting mixtures are largely chosen based on their physical and chemical properties, especially their improved capacity to concurrently provide roots with enough levels of oxygen and water. To make it easier for plant roots to extract water, the substrates should be porous and retain water with weak matric forces. The bulk density (BD) of substrates is often substantially lower than the BD of agricultural soils, ensuring adequate aeration and facilitating handling and shipping. This also results in increased total and air-filled porosities. In the event of disease contamination, a number of commercial soilless substrates, including stone wool, polyurethane, perlite, and vermiculite, are easily disinfectable between growth cycles. Additionally, they make it relatively simple to adjust the pH and the amount of nutrients in the root zone. The inherent restriction and confinement of root capacity by growing containers in soilless cultivation is a drawback. Since there are more roots per unit of medium capacity, the constrained root volume enhances root-to-root competition for water, oxygen, and nutrients. Jones has carefully examined descriptions of horticultural substrates and soilless culture management (2005). Due to the wide range of origins, physical and chemical features, and the regular introduction of new products, it is impossible to cover all substrate types that are currently on the market. Therefore, only the media that are typical and commonly used are described here. Substrates are divided into two types of materials in this chapter:

- Natural, unaltered materials such as sand, tuff, and pumice are categorized as "inorganic substrates."
- Materials that have been treated make up "synthetic substrates" (such as stone wool, perlite, vermiculite, expanded clay, zeolite, and foamed glass).

Polyurethane or phenolic resin are examples of substrates that are synthetic organic materials. The substrates' surface charge and chemical activity are important characteristics. As a result, substrates are classified as semi-inert (such as sand and stone wool) or active (such as peat and tuff) materials. These two groups' differences are quantitative rather than qualitative in character. Each substrate is described in terms of its manufacturing and origin as well as its overall suitability as a growing medium. For effective irrigation management, the physical properties including BD, water retention, and hydraulic conductivity are offered. The chemical properties that are necessary for effective fertilization and nutrient management are examined, including composition, stability as affected by pH, cation exchange capacity (CEC), and salinity. Since disease prevention is a key element in substrate reusability, information on substrate sterilization is supplied. Since environmental pollution is currently a major concern in extensive soilless agriculture, information on waste treatment is also provided.

Common Inorganic Substrates

Unmodified Natural Materials

Sand

The International Society of Soil Science defines sand as the coarse soil mineral component and classifies it into fine and coarse sand with dimensions of 0.02-0.2 mm and 0.2-2.0 mm, respectively. Due to its low cost and accessibility, pure sand is frequently employed in deserts and coastal plains. It is frequently used in growing beds that are positioned on the ground and protected from the soil below by a polyethylene covering. The particle sizes and distributions are frequently irregular in a natural deposit. The range of particle diameters determines the needed depth of the sand layer. To prevent water logging and inadequate aeration, a deeper layer is needed for finer sand. Sand is also a typical heavy ingredient in a variety of substrate combinations, acting as ballast for top-heavy plants and as a windbreak in outdoor nurseries. Sand has a high BD in comparison to other growth media components, with fine and coarse sand having BDs of 1.48 and 1.80 g/cm3, respectively. For fine and coarse sand, the total porosity is only 45 and 30 percent, respectively, while the volumetric water content at saturation is only 39 and 27 vol. percent, respectively. The small pore fraction of sand maintains essentially constant water volume over a matric potential range of 0 to 210 cm water for coarse sand and 0 to 220 cm water for fine sand due to the narrow pore-size distribution of sand. A further reduction in the matric potential causes the water content to drop dramatically (see figure below). Hysteresis is a very small phenomenon. According to its retention curve, aeration issues are to be anticipated when using fine sand in pots or beds. According to Bunt (1991), a fine sand bed's mean oxygen diffusion rate was 10-100 times lower than that of peat, perlite, redwood bark, and other combinations of those materials. At 5.1 and 7.1 cm/min, respectively, coarse and medium sand have moderately high saturated hydraulic conductivities. For matric potentials below 210 and 220 cm, respectively, the unsaturated hydraulic conductivity of coarse and medium sand falls off significantly.

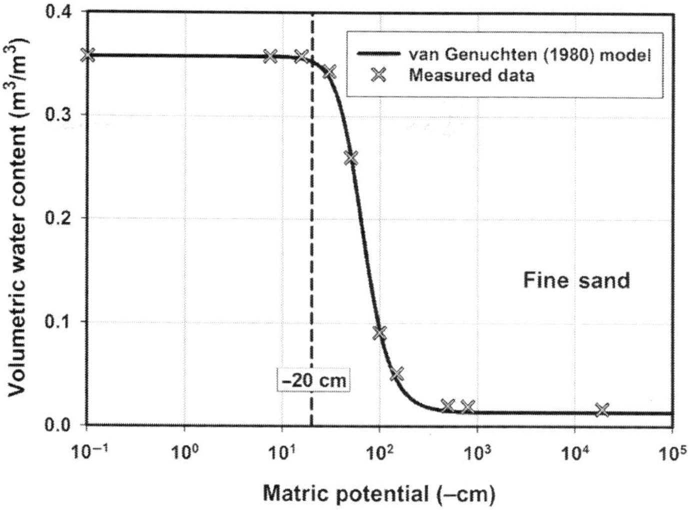

Figure 9: Water retention curve (WRC) for sand.

Raviv et al., (2019).

The most prevalent component of the sand fraction in soils is quartz (SiO2). It is the second-most prevalent mineral in the crust of the earth and is extremely weather-resistant (Drees et al., 1989). Quartz has a high particle density of 2.6-2.65 g/cm3 and a low specific surface of 1-10 m2/g. In the pH range of 2.5–9.0, quartz is a stable mineral with a low solubility of 3–7 mg Si/L. With very little Si being replaced by Al, Fe, and other minor elements, it is one of the cleanest minerals ever discovered. As a result, Quartz has a very low charge deficiency, which is a key factor in the physicochemical activity of other soil minerals. Sand can therefore be used as a diluent to more reactive components in potting soil because it is a chemically passive substrate. It is important to note that beach sand and sand used for deicing roads are salty and should not be used in horticulture because of this. Steam can be used to disinfect sand. Due to capillary condensation, thin layers can cause the pores to fill up quickly with water, disrupting the steaming process. Sand is particularly durable since using it as a growing medium does not produce any chemical or biological changes to it. There are no environmental contamination issues because sand waste can be utilized in building.

Tuff

Tuff, also known as volcanic scoria, volcanic cinder, lapillus, tezontle, pozzolana, and picon, is the common name for pyroclastic volcanic material, which is distinguished by high porosity and surface area. According to their silica content (silica percentage in the solid phase), volcanic rocks are categorized as rhyolitic (greater than 65), andesitic (50-65), and basaltic (less than 50). Rhyolitic lavas are created during eruption at very moderate temperatures (8000C–10000C), hence they mostly contain light elements like Si and Al while having low concentrations of Fe, Mn, Ca, and Mg. Rhyolitic rocks have a high viscosity and a pale tint. Basaltic lavas have a higher Fe, Mn, Ca, and Mg content than other types of lava because they are created at high temperatures (over 1000oC), which results in low viscosity and a black color. Andesitic lavas have a chemical makeup that is halfway between that of rhyolitic and basaltic lavas, are generated at intermediate temperatures, and have an intermediate tint. Since primary minerals cannot develop due to the rapid cooling of the magma during eruption, pyroclastic rocks are primarily made of vesicular volcanic glass. Tuff's mineralogical makeup and weathering processes have a large role in determining its physical and chemical characteristics. Tuffs come in different colors (black, red, and yellow), all of which were produced by the same volcano and have nearly the same chemical makeup. However, they have different primary mineral compositions and weathering stages, as well as unique surface charge properties, P adsorption capacities, and dissolution kinetics. The physical and chemical characteristics of tuff may also be impacted by the grinding and sieving procedures. Depending on where it comes from and how it was sieved or ground, tuff has a BD ranging from 0.8 to 1.5 g/cm3 and a total porosity between 60 and 80 percent. Tuffs are mostly made of amorphous materials and feature permanent and variable charge surfaces. During the growth season, P in particular may be absorbed or released by Tuff due to its considerable buffering capacity. Tuff's mineralogical makeup affects its chemical stability. Dissolution of secondary minerals, such as kaolinite and halloysite, is slower than that of volcanic glass, which dissolves quite quickly. As a result, nonweathered materials like black tuff, which contain a lot of volcanic glass, are unstable and dissolve rapidly in solutions with pH values below 6, as opposed to red and yellow tuff, which are more stable. Due to the toxicity of Al and Mn, exposing plants to black or even red tuff after equilibration with an acidic electrolyte ($pH < 5$) may be dangerous. Tough is a durable material that has a long lifespan. Due to the buildup of organic matter (OM) and low molecular weight fulvic acid, growing plants may potentially improve the chemical characteristics of tuff. For disinfestation, you can use chemical treatment, solarization, or steaming. After rose (Rosa 3 hybrida) cut flower production, tuff underwent deinfestation treatments, and the yields were better than those of unused tuff (Raviv et al., 1998). The solubilization of OM was strongly impacted by the disinfestation procedure, however, neither the solid phase nor the stability of the tuff was modified. The tuff surfaces later absorbed the soluble organic compounds released during disinfestation, modifying the surface charge (Silber and Raviv, 2001). Therefore, the disinfestation treatments improved food availability in addition to their beneficial impacts on pathogen populations.

Pumice

Like tuff, pumice is a byproduct of volcanic activity that often develops from silicic lavas that have a rhyolitic composition and are abundant in gases and volatiles. During volcanic eruptions, the rapid release of pressure causes gas to expand and create low-density materials made of extremely vesicular volcanic glass. Pumice is widespread in regions with a high rate of volcanic activity, including the Greek Islands, Iceland, the Portuguese Azores, Japan, New Zealand, Russia, Sicily, Turkey, and the United States. Quarry workers mine the raw material, which is then processed and sieved according to specifications. Its aggregate size has an impact on both its physical and chemical characteristics. Since the time of the Romans, pumice has been used as a filler in the paper and plastic industries as well as a lightweight aggregate in building, stonewashing for textiles, metal, wood, and glass cleaning and polishing, and filling for paper and plastic. Pumice's morphological characteristics resemble those of tuff extremely closely, but its chemical characteristics, particularly surface activity and charge, are very dissimilar. Pumice is an inert aluminosilicate material, whereas tuffs feature charge surfaces that are both permanent and variable and have the ability to buffer nutrients. Depending on its source and the sieving/grinding process, pumice is a lightweight aggregate with a low bulk density (BD) of 0.4–0.8 g/cm3 and a total porosity of 70–85 vol. percent. Pumice has big pores, therefore as the matric potential decreases (becomes more negative), so does the volumetric water content. Pumice has a comparatively limited water holding capacity when compared to stone wool, perlite, or organic substrates, which may restrict plant water and nutrient uptake, particularly in arid conditions. The air-filled porosity, total volumetric porosity, and readily available water (defined at a matric potential of: 0, 0.0 to 21.0 and 21.0 to 25.0 kPa, respectively). As matric potential declined from 0 to 210 kPa, the unsaturated hydraulic conductivity of Sicilian and Greek pumice decreased by nearly 4 and 6 orders of magnitude, respectively. Therefore, plants cultivated in pumice require very regular irrigation. Pumice is an inert aluminosilicate substance largely made of silica and A1-oxide, though it can also include salts, calcite, and metal oxides. Pumice, which mostly contains carbonate and metal impurities, has very little surface charge and negligible buffering capacity. Even at a pH of 2.5, the substance is still stable. Fresh pumice material should be applied with caution, though, as high Na concentrations could initially leak. Pumice is devoid of diseases and weeds and is biologically inert. It is reliable and nearly endlessly recyclable. Since it is a natural substance, it can be thrown away without harming the environment.

Processed Materials

Perlite

Glassy volcanic rock called perlite contains 2 to 5 percent mixed water and has a rhyolitic composition (Alkan, 2004; Kaufhold et al., 2014). About 70% of the world's known perlite resources are found in Turkey in the Aegean Sea. Heating the crushed and sieved material to 760°C–1100°C results in commercial perlite. The amount of energy used to produce perlite contributes to a significant environmental burden, according to a life cycle analysis (LCA) done to evaluate the environmental impacts of cherry tomato production using perlite as a substrate, but it also reduces the uncertainty involved in field production of this crop (Romero-Gamez et al., 2017). At high oven temperatures, the combined water in the perlite is transformed into gas, which causes the volume to expand to 4–20 times its initial volume and produce a light, porous material. Perlite is often used as a stand-alone growth media and in potting soil blends. It is made in different grades, with the most popular ones having diameters of 0-2.0 and 1.5-3.0 mm. The physical characteristics of the various grades vary. Bark was more cost-effective than perlite, which is a relatively pricey substrate. Similar to how perlite might be replaced with volcanic gravel. Perlite's higher price was, however, made up for in other instances by the higher yields it produced when compared to stone wool and pine bark. Expanded perlite is extremely light weight, with particle and BD of 0.7 and 0.1 g/cm3, respectively, thanks to enclosed air pockets. It is very porous and can hold three to four times as much water as it does.

Perlite is probably the most popular substrate for green roofs because of its extremely low BD, but one must take into account the substantial weight of the moist substrate. The medium ingredient with the highest air-filled porosity is likely perlite (Schindler et al., 2017). According to Bures et al. (1997a), the coarse fraction of expanded perlite (0.5-1.0 mm diameter) retains water at 210 kPa at a substantially higher rate than the fine fraction (0.25-0.50 mm diameter). The fact that the coarse and fine fractions can contain different amounts of water demonstrates that the coarse fractions can hold the majority of the water in internal pores. The amount of internal porosity alone cannot account for it, though. In comparison to sand and stone wool, the drop in water content with decreasing matric potential is minimal. Commercial perlite with a diameter of 0–4 mm had 13.6 and 36.5 vol. percent of both available and non-available water, respectively. Perlite's water retention curve (WRC) exhibits a modest amount of hysteresis. Regardless of its initial moisture level, perlite saturates fairly quickly. The saturation hydraulic conductivity is dependent on the size of the particles. Saturated hydraulic conductivity for commercial perlite with particles 50% smaller than 0.5 mm in diameter was 0.3 cm/min. The hydraulic conductivity was found to be reduced by two orders of magnitude as the matric potential dropped from 0 to 230 cm H2O. When compared to sand, this shift is mild. Perlite is commonly utilized as a component in blends based on organic substrates like compost because of its good physical properties. By doing this, the favorable biological characteristics of such media can be preserved for a considerable amount of time while avoiding compaction.

Although perlite has a pH between 7.0 and 7.5, it is neutral and devoid of any mineral nutrition or buffering ability. Perlite is a good medium for evaluating the utility of various kinds and quantities of nutrients because of its inertness and neutrality (Wik et al., 2006). A hazardous Al leak into the solution is possible when the pH is low. The table below provides the material's chemical composition as determined by Olympios (1992). Due to its production at extremely high temperatures, perlite is a sterile substance. Chemically, perlite is a substance with a long shelf life that is not significantly impacted by acids or microbes. Recycling perlite doesn't have an impact on the environment because it is an inert substance. Because of media compaction, salt buildup, and pest infestation, reusing perlite to grow subsequent crops without processing carries a possible risk. It is expensive to switch out used perlite for fresh media when growing successive crops, and in a market where prices are competitive, recovering the cost by charging more for the product might not be the best strategy.

To prevent pathogen contamination, spent perlite should be steam sterilized before a fresh crop is planted. However, to restore perlite's loose structure and to lower the accumulated salt content, this treatment necessitates pricey steam generators and might not be sufficient (Hanna, 2005). Hanna (2005) recycled perlite twice a year for commercial tomato production in bag (18.9 L) culture in an experiment to see if growing tomatoes (*Lycopersicon esculentum* Mill.) in cleaned and disinfected old perlite is more affordable than in new perlite and without any detrimental effect on yield. Used perlite was treated with hot water (13.25 L water/18.9 L perlite) at temperatures exceeding 93.3oC to drain excess salts and sanitize the medium after the roots from the previous crops were separated. This treatment significantly reduced salt content (Electrical Conductivity (EC), NO3-N, and K were reduced by 43 percent, 50 percent, and 47 percent, respectively) while raising media temperatures above those required to kill several fungi and nematodes. However, over the course of eight years, there was little change in the media's physical properties, such as the particle size distribution (PSD). Contrary to popular belief, perlite is not susceptible to mechanical compression and disintegration, as was discovered by Giuffrida et al. (2008) and Giuffrida and Consoli (2016). Cleaning and sanitizing spent perlite before recycling it resulted in heavier tomato fruit and a larger marketable yield than fresh perlite while saving 56% of the cost of replacing the medium. The combined effects of salt reduction, media cleaning, and the presence of an ideal level of nutrients were linked to the observed yield benefit of perlite recycling; it typically requires time to build up nutrients to an ideal level in new perlite. Because used perlite is not biological in nature and is physically and chemically stable, it can be cleaned and disinfected as needed and recycled for many years (Hanna, 2010). Additionally, Marfa et al. (1993) discovered that perlite keeps its physical characteristics for succeeding crops. The fact that recycled perlite and peat/perlite mix were more suppressive against *Fusarium oxysporum* f. *sp. radicis lycopersici* than the brand-new, unused media is an extremely significant finding. The growing bags inside the greenhouse can be solarized, which is another effective method of cleaning the perlite (Perez-Hernandez et al., 2017).

Table 1: Chemical Composition of Vermiculite

Oxide	SiO_2	Al_2O_3	MgO	Fe_2O_3
Percent	20-25	5-10	35-140	32-35

Source: Raviv et al., (2019).

Vermiculite

Vermiculite's primary component is a naturally occurring clay mineral with layers and water filling the spaces between the layers. Similar to how perlite is made, expanded vermiculite is created by heating the ground and sieved material to 1000°C. In the oven, the water is heated to a high temperature and turned into vapor, increasing interlayer space. Exfoliation is the word for this procedure. Expanded vermiculite is made up of light, accordion-shaped granules with excellent porosity as a result. Vermiculite is a component of potting soil mixtures and is used as a growth medium. While coarse grades are widely utilized as rooting media, fine grades are mostly employed as mulch in the production of transplants. The most popular sizes and grades of expanded vermiculite are 0-2, 2-4, and 4-8 mm in diameter. With a particle density of 0.9 g/cm3 and a BD range of 0.07 to 0.1 g/cm3 for the coarse and fine grades, respectively, all grades are exceptionally light weight. It has a high porosity, absorbs water quickly, and can hold three to four times its weight in liquid. Vermiculite is a clay that has a pH between 7.0 and 7.5 with a low EC. For every octahedral sheet, there are two tetrahedral sheets in this form of clay. It also possesses a persistent negative charge, much like the raw material. As a result, it has a pH and cation buffering capacity and a CEC value of 15–21 cmol/kg. Due to its large surface area and a few positive charge sites that are situated at the borders of the clay tactoids, it also absorbs ions like phosphate. This may be the cause of why vermiculite performed better than perlite in comparison experiments. Table 1 above provides the material's chemical make-up, represented as oxides (percent). There is a chance that poisonous Al will release into solution when the pH is low. The major adsorbed cation and an essential part of the mineral structure is magnesium. Because it is created at extremely high temperatures, vermiculite is a sterile substance. It can't be steam sterilized though because heating causes it to break down. A structure made of inflated vermiculite is quickly destroyed. As a result, it cannot be used for an extended period of time. It is delicate to mechanical compression, which could cause particles to get pulverized. Vermiculite disposal does not pose a threat to the environment. In some vermiculite mines, asbestos was formerly discovered. Since all of those mines were shut down, vermiculite is now regarded as a secure growth medium.

Expanded clay granules

Expanded clay is a granular substance with an internal and external crystalline structure. It is created by shaping the clay into pellets and baking them in rotary kilns at 1200oC, where gas is released and the clay expands. Low soluble salt content in the raw material is crucial in order to avoid adding additional ingredients like lime to the process. Otherwise, during cultivation, salt might be leached. In horticulture, grade sizes with a diameter of 4-8 or 4-16 mm are used. Since 1936, expanded clay granules have been utilized in horticulture. They are employed in hydroculture and in the construction of plant containers. Expanded clay granules have a BD of 0.28 to 0.63 g/cm3 and a water content of 11 to 24 vol% at a matric potential of 210 cm H2O. They are light in weight. High BD was reported by Wever and van Leeuwen (1995), with 1.2 g/cm3 and 53 percent porosity. They discovered that as the matric potential decreased from 23 to 250 cm H2O, the water content decreased from 48 to 44 percent, meaning that there is only 4 percent readily available water. The material, on the other hand, has a lot of enclosed air. After five years of intensive cropping, it has been demonstrated that there were no changes in physical characteristics. When starting from a dry state, hydration of the substrate is relatively slow; however, if it is started at a matric potential of 2100 cm, it is quick. The WRC doesn't show any minor hysteresis. The hydraulic conductivity of expanded clay granules is unknown, but they are comparatively resistant to compression. Granules of expanded clay have a pH of roughly 7.0, making them neutral. If low salt clay was used and no salts were added during baking, the EC is low; if not, washing is necessary. It is categorized as an inert substance that is incapable of cation exchange or buffering. However, Meinken (1997) showed that nutrients can build up within the granules as a result of diffusion and then be released back into the solution. Due to their high-temperature production, expanded clay granules are a sterile product. They may be sanitized and washed after usage without suffering any negative consequences. Granules of expanded clay have a long shelf life and are quite stable. Construction companies can use waste materials.

Zeolite

Zeolites are limitless, three-dimensional crystal forms in hydrated crystalline aluminosilicates with alkali and alkaline cations. Zeolites can also originate from nonvolcanic materials in marine deposits or aquatic environments. Zeolites are typically created by the metamorphism of volcanic rocks. Zeolites are widely utilized in agriculture and many other industries for the removal of pollutants from waste and drinking water due to their ion exchange, adsorption, hydration/dehydration, and catalytic characteristics. Synthetic minerals are created to improve the chemical and physical characteristics of natural zeolite. Worldwide, 508 different varieties of synthetic zeolites with enhanced physicochemical qualities are produced, according to Sherman (1999). In soil remediation, zeolites (mostly clinoptilolite) are employed to absorb heavy metals or radioactive waste. They are employed in agriculture as soil supplements to increase water availability, decrease nitrate contamination and losses, and provide P, K, and NH4 nutrients in infertile soils and substrates. Zeolite can be used as a sophisticated example of a growth medium by pre-charging it with aluminum to create blue hydrangea. Zeolites have been used in agriculture in several ways, according to Pond and Mumpton (1984). Zeolites, primarily clinoptilolite, are currently a common component of growth media, particularly in Mediterranean nations like Greece and Turkey. Zeolites have BDs that are reasonably high (1.9-2.3 g/cm3) and exceptionally high CEC values (220460 cmol/kg), hence using them as a single component growth substrate is not advised. However, zeolites are commonly utilized for flower and vegetable cultivation all over the world (Bulgaria, China, Cuba, Italy, Japan, Jersey, Korea, Russia, United States, and Yugoslavia) in mixed substrates that include contain inorganic elements (sand and perlite) as well as organic ones (peat and compost). For numerous crops, including gerbera (*Gerbera jamesonii* Bolus ex Hooker f.), lettuce, sweet pepper (*Capsicum annum* L.), tomato, and carnation (*Dianthus caryophyllus* L.), zeolite has been used experimentally as a single growth substrate (*Lactuca sativa* L.). Only the physical and chemical characteristics of clinoptilolite, the main zeolite material utilized in agriculture, are mentioned.

Clinoptilolite has a saturation water content of 34% vol.% and a particle density of 2-2.1 g/cm3. However, the chemical and physical properties are altered by grinding and sieving. Turkish zeolites (clinoptilolite) with different aggregate sizes had physical characteristics that were nearly equal to basaltic and rhyolitic tuff with comparable particle sizes. Zeolite can be used in roof gardens because its BD is lower than that of soil, an application that has grown in popularity over the past ten years. Its relatively high water retention capacity makes storm-water management effective, which is crucial in urban settings. Although relatively stable, clinoptilolite dissolves at pH 2 or lower. Clinoptilolite has a CEC of 220 cmol/kg and a unit cell formula of [(Na3K3) (A16Si30O72)]24H2O. (Ming and Mumpton, 1989). Zeolites can potentially be reused because they are not sensitive to mechanical compression and handling does not alter their physical characteristics. For the purpose of growing sweet pepper over a four-year period, Harland et al. (1999) investigated zeolite (clinoptilolite) as a fresh and recycled substrate in closed (recirculating) and open (nonrecirculating) fertigation systems (three recycling cycles). Over the course of the trial, there was no decrease in yield or quality attributable to the reuse of zeolite substrate. In the recycling system, the plants were removed and the roots were sieved from the substrate at the end of each cropping period. The substrate was then steam-sterilized at 90oC for 3 hours before being re-bagged and used as substrate for the following crop.

Foamed Glass

Glass waste is used to create foamed glass aggregates. The first step in the manufacturing process is to crush and grind recycled glass into a very fine powder, which is then combined with calcium carbonate (2 percent on a mass basis) and additional foaming agents. In a foaming furnace, the fluid is heated to around 850oC before being put into molds. The molten mixture expands as a result of the foaming agents' evaporation, resulting in a network of tiny pores. The foamed glass slabs are crushed, tumbled, and sieved into different aggregate sizes for uses in horticulture and aquaculture once they have cooled to room temperature. Foamed glass is also used in the construction industry as a fire-resistant insulating material, as a frost-heave insulation, and as a lightweight bulk fill material for roads. Growstone (Growstone, LLC, Albuquerque, NM, USA) sells foamed glass aggregates for use in horticulture. They come in a variety of sizes and are combined with coconut coir. Tomatoes planted in Growstone aggregate growth bags at the University of Arizona Controlled Environment Agriculture Center produced equivalent yields and plant development to tomatoes grown in stone wool, according to preliminary greenhouse studies (Chen Lopez, 2011). The sugar content of the fruits was much higher in Growstone than in stone wool due to the enhanced aeration qualities of the Growstone aggregates. Tomato output using Growstone aggregates surpassed stone wool by 4% in a similar trial at Wageningen University and Research Centre (Costa, unpublished data). Stone wool had a greater incidence of blossom end rot. Foamed glass aggregates have a BD of between 0.15 and 0.25 g/cm3 and an accessible specific surface area of roughly 15 m2/g, depending on the aggregate size distribution. They are also quite light. With a portion of small intra-aggregated holes and a portion of larger interaggregate pores, they display a typical bimodal pore size distribution. The interaggregate porosity accounts about 70% of the total pore volume in foamed glass aggregates with a 0.85 cm mean diameter. According to the two different pore fractions, the WRC displays a bimodal shape (see Figure below). In contrast to stone-wool, foamed glass aggregates have excellent aeration qualities that reduce the risk of pythium. After reaching saturation, the interaggregate pores rapidly drain over a very small matric potential range. The water that is accessible to the roots is stored in the intra-aggregate pores, which release water at a much slower rate.

It's interesting to note that by simply changing the aggregate sizes, the aeration/water holding properties can be adjusted and tailored to a particular crop. Foamed glass aggregates with a mean diameter of 0.85 cm had a saturated hydraulic conductivity of 124 cm/min. Chamindu Deepagoda et al. report the physical and hydraulic characteristics of a 1:2 (v:v) foamed glass aggregate/coconut coir mixture (2013a). Although there isn't much information available about the chemical properties of foamed glass aggregates, it is advised to rinse the aggregates before transplanting them to lower the pH of the substrate. The gradual release of monosilicic acid from the foamed glass matrix, which builds up in plant leaves and shoots, is an intriguing chemical aspect. Increased silicon concentrations in plant leaves have been shown to have positive effects, including resistance to Mg toxicity, fungal diseases like Pythium, and perforating insects like leaf miners. New foamed glass aggregates offer a root environment that is initially less prone to diseases since they are a sterile product. The material structure of foamed glass aggregates is not adversely affected by steam sterilization, indicating that the aggregates can be reused without losing strength.

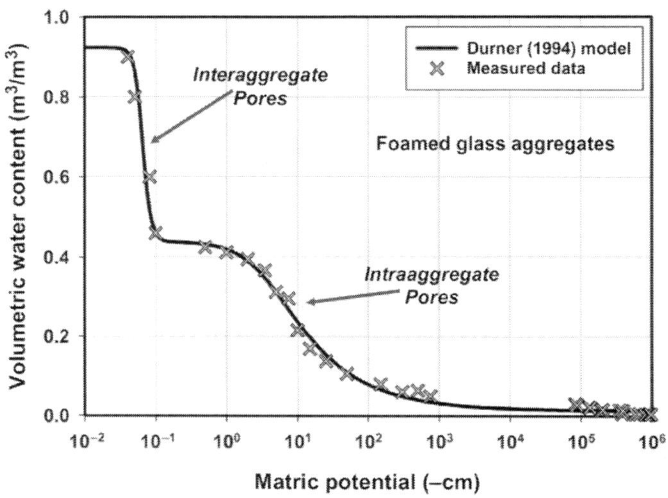

Figure 10: WRC for foamed glass aggregates.

Source: Raviv et al., (2019).

The water retained in the inter- and intra-aggregate pore fractions is reflected in the bimodal form. The fitted Durner (1994) dual-porosity water retention model is represented by the solid line. Water retention curve, or WRC.

Mineral Wool

Stone and glass wool

Production, place of origin, and details— Mineral wool is a thin, synthetic material that was initially created for thermal and acoustic insulation in the building sector. Products made from stone wool were initially created for a number of uses. In 1969, Denmark pioneered the use of modified varieties of stone wool as a foundation for horticulture (Smith, 1987). The GRODAN Group arose as a result of several business changes, and by the 1980s, the term "rockwool," the brand name for its flagship product, had come to mean stone wool. While available in granulated form as part of potting mixtures, stone wool is most commonly employed in horticulture in the shape of slabs or blocks of connected fibers. Three natural raw materials, 60 percent diabase (a type of basalt rock, dolerite), 20 percent limestone, and 20 percent coke, are heated to create stone wool. High temperatures are needed to fuse these materials, with coke serving as fuel in a blast furnace that uses forced air to boost the temperature to 1600 °C. The molten mixture is then spun rapidly into thin fibers with a diameter of roughly 0.5 mm, which are cooled by an air stream. To bind the fibers together and reduce the material's inherent hydrophobicity, chemicals like phenolic resin and wetting agents are heated with the fibers. After that, it is shaped into slabs or blocks of various sizes. A sizeable portion of the overall energy needed for greenhouse tomato and other crop production is used in the manufacture of stone wool. The alleged detrimental impact on the health of stone wool is still another significant downside (Kudo et al., 2009). Quartz sand is melted at 1200°C in electric furnaces to create glass wool. Similar to how stone wool is made, glass wool is made into slabs and fibers. Both stone wool and glass wool (mineral wool) are often sanitary, manageable, and perform consistently. Today, stone wool is one of the main greenhouse substrates used in practically all regions where greenhouse vegetables are grown, including The Netherlands, Belgium, Germany, Denmark, and other Western European nations (Canada, United States, Netherlands, and other countries). Horticultural crops can grow well in it because gardeners can easily control the amount of nutrients and the ratio of water to air in the root zone. However, due to the tiny volume of the common slabs and the lack of a buffering capacity for nutrients, pH, and water, it is "unforgiving" to management faults.

Depending on the manufacturer, stone wool has a low BD of approximately 0.05–0.2 g/cm3 and a porous volume of 92–98 percent. High water retention is observed at matric potentials above 21 kPa, and as the matric potential lowers, the water content rapidly diminishes until almost no water is kept at potentials below 25 kPa. As a result, after irrigation and free drainage, the slab has a poor water buffer capacity and a high gradient in water content from top to bottom. At a height of 1 cm above the base, there is only 4% air volume, whereas there is 100% dryness in the top layers. For the best water to air ratio, a slab should be 7.5–10 cm high. Van der Gaag and Wever (2005) suggested using thicker slabs while growing cucumbers, as the high water content of stone wool in shallow slabs makes it more favorable to *Pythium aphanidermatum*, the cause of Pythium root and crown rot. During the first wetting/drying cycle, pronounced hysteresis was seen, but it subsided in subsequent cycles. At saturation, the hydraulic conductivity is relatively high (5–20 cm/min), but it rapidly drops as the matric potential drops (da Silva et al., 1995). If irrigation is put off, such decreases in hydraulic conductivity may result in inefficient absorption of water and nutrients as well as the quick onset of water stress. When stone wool is squeezed under pressure, it retains its previous height after relaxing. However, after use, stone wool softens and loses some of its elastic properties. Glass wool is incredibly light and has high water and air capacities. In contrast to stone wool, glass wool's fiber diameter can change, which has an impact on how much water it can hold. It is feasible to get better water content distribution across the slab's height by having finer fibers in the higher portion of the slab than in the lower portion. The table below lists the stone wool's chemical makeup in terms of oxides (percent). Stone wool's primary chemical property is its relative inertness, with the exception of a few small pH-related reactions. The initial pH of commercial stone wool used to be very high (7.0–8.0), and values as high as 9.5 have been recorded. However, the industrial standard nowadays is values of 6.6-6.5. Low pH values should be avoided since stone wool dissolves below pH 5.0. Stone wool by itself has no impact on crop nutrition, hence irrigation water is required to provide all nutrients. However, it was shown that several plants and rose rootstocks can absorb a significant amount of Fe from stone wool (Sonneveld and Voogt, 1985). Additionally, according to Rupp and Dudley (1989), DTPA may be used to extract significant concentrations of Fe, Mn, Cu, and Zn from a variety of stone wool products.

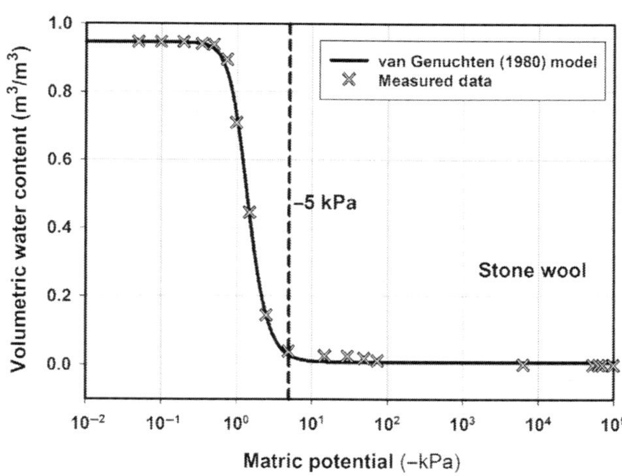

Figure 11: Water retention curve for stone wool.

Source: Raviv et al., (2019).

New stone wool offers a root environment that is initially less vulnerable to diseases because it is a sterile product, especially in closed-loop fertigation systems. Prior to usage, stone wool can be steamed. Stone wool is often thrown after one or two crop cycles, resulting in a significant amount of waste (about 125 m3/ha of plant productivity). However, stone wool has occasionally been utilized for three crop cycles with success. The very large waste volume is a significant environmental concern because stone wool does not naturally disintegrate. In many nations, it is no longer sustainable to dispose of spent horticultural stone wool in landfills. Different techniques for recycling stone wool have been developed over the past 20 years. Manufacturers of stone-wool, notably GRODAN A/S (Denmark) and GRODAN BNF, have done and continue to conduct research on the reuse of stone-wool (The Netherlands). Several recommendations for reusing horticultural stone wool were included in the GRODAN studies summarized by Bussell and McKennie (2004), including recycling by a stone wool factory, soil amendments (to improve the poor physical properties of clay soils), mixing with other substrates (like peat and compost), brick production, landfill cover, mining cover/recultivation, and mushroom casing. However, the preferred method of use for sterilized stone wool is direct reuse. Sadly, sterile inoculated stone wool increased disease infections. On the other hand, the capacity of *P. aphanidermatum* zoospores to infect was removed by composting used stone wool that had been inoculated with yard and leaf waste. In both the United Kingdom and the Netherlands, where nearly all of the used stone wool from crop production is collected and processed at sizable facilities, reuse of stone and glass wool as a raw material for both horticultural and insulation applications as well as in brick production is now widely practiced. Stone wool is frequently disposed of in landfills in locations where this form of recycling is not practical. However, a study on several substrates for green roofs found that stone wool should be avoided in order to create roof systems that are environmentally friendly.

Table 2: Chemical Composition of Stone Wool

Oxide	SiO_2	Al_2O_3	CaO	MgO	Fe_2O_3	Na_2O	K_2O	MnO	TiO_2
Percent	47	14	16	10	8	1	1	1	1

Source: Raviv et al., (2019).

Incorporating discarded stone wool slabs (as finely chopped particles) into peat- or compost-based media for horticultural production is one of the possible uses for these materials; doing so offers the chance to recycle the stone wool and cut down on the usage of nonrenewable resources like peat. After the plastic sleeves were removed, used stone wool slabs were tested in a pilot study in New Zealand to see if they could be used as mulch around avocado plants. The mulch effectively stifled weed development for two years without harming the growth of avocado trees (Bussell and McKennie, 2004). The inability to steam sterilize stone wool slabs completely between crops, their propensity to break when moved, the presence of varying amounts of root matter in used slabs that makes water content and EC control much more challenging than in new slabs are the main obstacles to direct reuse of stone wool slabs for growing succeeding crop(s) (Bussell and McKennie, 2004).

Synthetic Organic Media

Polyurethane

The base ingredient needed to create the numerous foams, slabs, and granules utilized in the construction and furniture sectors is mineral oil. Glycol and mineral oil products like di-isocyanate can be combined to create polyurethane. The end product is a polymer containing extra di-isocyanate groups that, when they interact with water, produce CO2 and cause the polymer to foam. Furniture manufacturers employ the resulting foams, and growth substrates are made from the cutting remains. They are reduced to granules, which combine with additives to produce a slab when put under pressure. During the procedure, steam at 140oC is blown in. In Belgium and the Netherlands, foam is employed as a growing medium for flowers and vegetables. The granules are frequently included in combinations as an ingredient. Perlite is thought to be a suitable replacement for polyurethane in potting mixtures (Cole and Dunn, 2002). Similar to stone-wool, perlite can also be replaced with phenolic foam (Han et al., 2014). With an extremely low particle density and BD of 1.19 and 0.078 g/cm3, respectively, polyurethane is a relatively light substance. As a result, at 95%, the porosity is extremely high. Polyurethane holds a lot of air and very little water at low matric potentials, similar to stone wool, because the majority of the pores are rather big. Only 2% of the total volume of the matric potential between 210 and 2100 cm H2O is available water. Information on its hydraulic conductivity properties is not readily available. Polyurethane has a low EC and pH of 6, making it chemically inert. Except for Fe, Zn, and B, it does not release or contain any significant nutrients. The polyurethane foam used to contain dangerous organic compounds; however, this issue has been resolved.

Polyurethane cannot be composted after use because it is resistant to acids and is not broken down by microbial enzymes. It is a clean substance that can be used again during growing cycles between crops. Waste can either be burned or recycled to make polyurethane slabs for building. Polyurethane is simple to handle and can be reused for up to 10-15 years. Mechanically removing the surface roots and pressing the slabs or mats dry make it easier for steam to penetrate through, improving sterilizing. During four years of cultivation, Benoit and Ceustermans (1995) demonstrated that roots left in polyurethane mats enhanced water retention at 21 kPa from 8 to 16 percent, while air volume fell from 85 to 74 percent, indicating a shift in the pore size distribution to smaller pores. In a 10-year trial, Benoit and Ceustermans (1995) found that water retention at 21.0, 21.7, and 22.0 kPa increased in recycled polyurethane substrate while air volume decreased over time. After 10 years and eight steam treatments—one every two years—the polyurethane substrate's air/water retention properties were comparable to those of new stone wool. However, because of its low density, the substrate was prone to compression without being able to regain its previous volume, which could be problematic if it were to be regularly reused and steam-treated. The reuse of polyurethane substrate during tomato production did not have an impact on overall production or earliest harvest (Benoit and Ceustermans, 1994). When growing cucumbers (*Cucumis sativum* L.), the same polyurethane slab was used repeatedly over two years to produce yields that were marketable and comparable to those obtained with fresh perlite or organic substrate each year (the best yielding media). However, polyurethane required more irrigation water, which increased the leaching percentage. According to Hardgrave (1995), the productivity and fruit quality of cucumbers grown on previously used steam-sterilized polyurethane foam slabs were on par with or better than those grown on new polyurethane foam slabs or stone wool. These positive effects on crop performance were attributed to the recycled slabs' improved water-holding capacity as a result of the inclusion of root materials. Additionally, tests revealed that a polyurethane foam slab suffered significantly less physical harm than a stone wool slab when steaming between two succeeding crops since it was stronger and more flexible. Polyurethane is regarded as an advantageous substrate economically, agronomically, and environmentally because to labor savings and its durability with repeated application. According to Benoit and Ceustermans (1995), the cost of steam sterilization was only half that of recycling stone wool slabs, and the greater initial purchase price of polyurethane could be repaid over a minimum of 10 years.

Polystyrene

Expanded polystyrene foam flakes or beads are leftovers from the production of polystyrene. To increase aeration and drainage, they were occasionally added to substrate mixes. One of the most well-known synthetic polymers is polystyrene, a mass-produced plastic that is widely used. Other synthetic polymers include polyethylene, polypropylene, and polyester. Styrene, a liquid hydrocarbon that was created in the late nineteenth century from styrax balsam, a substance that comes from a tree in Asia Minor known as the oriental sweet gum, is the fundamental component of polystyrene (*Liquidambar orientalis* Mill.). In 1950, polystyrene was developed primarily as a thermal insulator. Currently, any grade, class, or kind of thermoplastic polymer, copolymer, interpolymer, alloy, or blend that contains primarily polymerized styrene is referred to as polystyrene. Suction polymerization is the process used to create polystyrene. Ethylene ($C2H4$) and benzene undergo a two-step reaction to produce styrene ($C8H8$), which is then first produced from petroleum or natural gas ($C6H6$). In order to create ethylbenzene ($C8H8$), ethylene first combines with benzene ($C6H6$) in the presence of a catalyst, such as aluminum chloride. Styrene is then created when the hydrogen is removed from styrene at temperatures between 600 and 650 degrees Celsius. After that, styrene is polymerized either through heat or an initiator like benzoyl peroxide. By stopping the polymerization process at a certain point with polymerization inhibitors (also known as terminating agents), such as oxygen, sulfur, or quinol, the degree of polymerization can be controlled. Then the droplets of polystyrene are suspended in water to form low density, loosely attached cells. The water is then mixed with a suspension agent, such as specially precipitated barium sulfate or copolymers of acrylic- and methacrylic-acid and their esters (organic product formed by the reaction of an acid and an alcohol). The droplets are held up by the viscous suspension agents, which prevent them from adhering to one another. Small and tough polystyrene beads are created during suspension polymerization. Special blowing agents, such as propane, pentane, methylene chloride, and chlorofluorocarbons (CFCs), are employed to enlarge them. Following cleaning, any beads with unusual sizes are taken out of the polystyrene beads that were created. Many manufacturers used polystyrene in growing media during the 1980s. It has been demonstrated that polystyrene did not offer any advantageous characteristics that could not have been provided more effectively by other materials. The only effect of polystyrene beads is to reduce the effective root zone volume that is conducive to root development because the air inside the beads is isolated from the root zone air phase and because the beads don't have any distinctive water-holding or ion-exchange capacities. The environmental contamination around the nurseries caused by polystyrene beads that had flown or washed out of containers and were clogging drains was the major issue for almost all farmers. Since then, almost all growers have stopped using polystyrene.

Perlite shares similar physical traits with polystyrene beads. They have a 1.03 g/cm3 particle density. 98–99% of the air inside the polystyrene beads is trapped. They have an incredibly low bulk density (BD) of about 0.024 g/cm3. Polystyrene has very no water-holding capacity; yet, when added to growth mixtures, it improves aeration and drainage, lowers the BD, and increases drainage. Matkin (2005) measured a maximum volumetric water retention of 27 percent for the polystyrene mix as compared to 42 percent for the perlite mix in a study comparing the addition of 0.38 m3 of polystyrene beads or perlite to a growing mix (0.38 m3 sphagnum peat moss 12.27 kg, calcitic lime 11.36 kg, and dolomitic lime). Chemically, polystyrene beads are made up of 94–96 percent polystyrene, 4–6 percent blowing agent, 0,4–0,5% water, and 0,4–6% styrene monomer. They lack CEC or anion exchange capability, are nutrient-free, and are chemically inert. It has a neutral pH. Because of their huge particle size and hydrophobic makeup, polystyrene beads are extremely resistant to microbial degradation. High temperatures and specific chemical disinfectants may break them down. Additionally, they are vulnerable to liquid petroleum products like gasoline. They shouldn't be cooked with steam because it could allow the beads to rise to the surface of the growth media, where their light weight would encourage erosion by wind and water. They may be burned safely, producing only carbon dioxide and water. Because polystyrene includes CFCs, which are inert and safe for humans and the environment when released, handling polystyrene should be done with extra caution. But it took scientists a while to recognize that CFCs contribute to the ozone layer's deterioration after they were first used. As a result, it is no longer advised to use polystyrene that has been blended with other components for recycling.

Polyester Fleece

Polyester fleece is a soft, fuzzy fabric that is frequently used for clothing, blankets, and other uses that call for a warm, wool-like material. Due to its two-sided pile construction, the fabric has a layer of chopped fibers on both its front and back surfaces, making it resemble corduroy or velvet. Pure polyester strands are combined and thermally bonded to create polyester fleece. Polyester is created by heating terephthalic acid, a petroleum derivative, and ethylene glycol, a petroleum derivative also known as antifreeze, at temperatures between 150 and 210 degrees Celsius. Dihydroxydiethyl terephthalate, the product of this initial reaction, is then fed into an autoclave and heated under pressure to a temperature of around 280 °C. The substance turns into a polymer known as polyethylene terephthalate at this temperature. The polymer solidifies into a thick, viscous liquid as it cools. After then, this liquid is ejected through a nozzle resembling a showerhead (tiny holes in a metal disk called a spinneret). The streams of liquid polymer dry and harden as they come into touch with air. This polymer was known as terylene in England; Du Pont obtained exclusive US rights to the material in 1946 and gave it the name polyester. The polymer polyethylene terephthalate can be produced into the plastic typically used for soda bottles even if it is not extruded into fibers. Polyester fiber manufactured from recycled soda bottles was developed in the 1980s as a result of the growing interest in recycling plastics. Many of the polyester fleece clothing options available now combine recycled and virgin polyester in their construction. As a horticultural medium, plastic foil-covered slabs of polyester fleece are employed. When opposed to stone wool, which requires 8.8-12.5 t/ha of substrate, polyester fleece can be produced with as little as 1.7–3.6 t/ha of substrate needed.

According to Schroeder and Forester (2000), polyester fleece has a water-air-solid ratio of 59:39:2, which is ideal for horticultural crop growth at a matric potential of 20.5 kPa. A polyester fleece slab has a pore volume of 98 percent, which is extremely similar to a slab of stone wool. It is a media that is quite light. The BD can vary significantly depending on the fleece's quality, the fibers' structure (compactness), and the height of the slabs. Schroeder and Forester (2000) employed polyester slabs with BDs of 0.045, 0.045, and 0.026 g/cm3 and three different heights (2, 4, and 8 cm) at the same length (100 cm) and width (20 cm). In comparison to stone wool slabs, polyester fleece slabs have much less water retention in the matric-potential range of 21 to 23 kPa. According to Schroeder and Forester (2000), stone wool's water retention stayed at about 70% for a fall in matric potential from 0 to 21.0 kPa, but polyester fleece's water-holding capacity dropped dramatically from 98 to less than 10 percent. Because the polyester fleece fibers are pH neutral and pH-stable from pH 2 to 11, it is simple to regulate the amount of nutrients in the rhizosphere. The CEC is nearly zero cmol/kg, making the material completely chemically inert. Because fleece can withstand temperatures of up to 160°C, it can be sterilized with steam after usage and used again for several more crops. The performance of repurposed fleece slabs is yet unknown because of the dearth of research findings. Fleece can be thrown away in landfills or utilized to make new slabs. It might be repurposed as a soil amendment (to enhance the physical characteristics of clay soils) or it might be used to substrate mixtures to balance out the retention of water and air. The fact that dried fleece waste can be burned is a significant benefit of fleece products.

ORGANIC SOILLESS MEDIA COMPONENTS

Introduction

When opposed to production in soil, soilless cultivation of plants in containers requires a constrained root system and a smaller root zone volume. In the past, dirt that was readily available in the area was utilized to cultivate plants in pots and containers. In addition to enhancing airflow and water retention, organic elements like manures and composts were frequently utilized to complement the inorganic base materials. These resources also provided nutritional advantages. In periodicals and other publications from the 19th century, there are several descriptions of mixtures of soils with organic elements, particularly peat and manures, and there are even a few reports of growing media made entirely of organic materials (Anon, 1830). Lawrence and Newell (1939) recommended adding 25% organic matter in the form of moss (sphagnum) peat in their scientific studies on the development of standardized growing media with constant pH, absence of pathogens, pests, and weeds, and without nutritional imbalances (the so-called John Innes composts). However, because their media needed the acquisition and sterilization of good-quality loam soil, professional growers found them less and less popular after 1950 as a result of the difficulty in locating such loam and the high expense of sterilization. Growing media, primarily based on peat, were created in the 1950s in the United States at the University of California. Penningsfeld and Kurzmann's 1966 research in Germany revealed that peat alone might serve as the foundation for growing media. The use of peat in horticultural growth media then grew quickly. With the notable exception of mineral wool substrates used for glasshouse crops like tomatoes, cucumbers, and peppers, the most widely utilized growing media are made mostly from peat and other organic materials as of the early 21st century. Schmilewski's data from 2017 show how peat predominates in organic growth media produced in Europe. In some nations, peat use and production have fallen. For instance, in the UK, peat use in horticulture decreased from 2.5 M m3 in 1997 to 2.0 M m3 in 2015: In fact, the government of the United Kingdom has decided to gradually phase out peat in growing media for retail markets by 2020 and professional markets by 2030. As an example, in Latvia, peat production for growth media increased from 0.65 million cubic meters in 1997 to 2.85 million cubic meters in 2009. In other countries, Canada produced roughly 17 M m3 of peat for horticulture use in 2015 (CSPMA, 2017), while Russia may produce comparable or somewhat higher volumes. Canada has about 1 M km2 of peat reserves. Peat is a semi-finite resource that regenerates relatively slowly after extraction and serves as a carbon store. Since the late 1980s, environmental advocacy groups have pushed for a reduction in peat harvesting and its replacement with other organic materials that are thought to be renewable.

The most prevalent of the latter is coir, whose use has grown significantly around the globe since the turn of the century. Other, frequently nearby, organic resources such as aged or decomposed bark, wood fibers, rice hulls, and composted materials are added to growing substrate. In order to improve the physical and, in certain circumstances, physicochemical features of organic growing media, inorganic materials including perlite, vermiculite, expanded clay, and pumice are routinely used. In Chapter 7, "Inorganic and Synthetic Organic Components of Soilless Culture and Potting Mixtures," inorganic media components are covered. There is growing public pressure to recycle industrial waste. High-quality horticultural plants can be produced using alternative organic substrates based on recycled materials that are well defined and rectified by effective mixing with inorganic components. This helps to lessen the overuse of natural peatlands. As a result, the use of peat-reduced media is a trend that is likely to persist. In this regard, just like with peat, new organic growing media components must be researched scientifically and technically while assessing the environmental and social impact associated with their widespread use in horticultural media mixes. Reusing substrates, if necessary, following disinfestation may become a top priority as the supply of resources like peat moss becomes more scarce. Responsible techniques for disposing of used media need to be established if this turns out not to be a practical choice. This chapter's foundation is based on the most significant organic compounds' individual and combined properties. Physical and structural traits, physicochemical parameters, nutritional status, and, occasionally, biological/microbiological characteristics are mentioned. Also taken into account are their stability, sustainability, and intended purpose.

Common Organic Substrates

Peat

In wet habitats where biomass accumulates in conditions of low pH and low oxygen content that are not favourable to microbial activity, peat is generated through the gradual decomposition of mosses, reeds, and sedges. The peatlands of greatest interest to horticulturists are found in the higher middle latitudes (50-60N) of the Northern Hemisphere, especially in Canada, Russia, and nations bordering the Baltic Sea. Peatlands can be found in a wide variety of temperatures. After the ice sheets receded 14,000–11,000 years ago, the postglacial epoch is when these peat deposits were formed. Peat bogs can be categorized as either topographic (basin bogs), where new peat deposition is influenced by topography and the groundwater table, or ombrogenous (raised and blanket bogs), where the growth of the peat layer is regulated by rainfall. Peat bogs first developed in postglacial lakes or regionally damp depressions. Peat bogs are therefore created by the incomplete breakdown of aquatic vegetation, followed by the growth of sedges and reeds (Carex and Phragmites spp.). Because of the buildup of plant detritus and the lack of breakdown of plant materials that have died, the bog slowly enlarges, allowing the lake's expanding vegetation to encroach onto it as the water is displaced upward. Higher water tables result from this, which also makes it possible for peat-forming plants to spread past the boundaries of the original lake. The groundwater influence decreases as peat formation proceeds, and nutrient availability is dependent more and more on rainfall. If this is enough, a shift to a raised bog occurs, when species of moss like Sphagnum and Hypnum take center stage. The International Peatland Society has suggested the following simplified system for classifying peat based on its botanical makeup, level of decomposition, and nutrient status:

Table 3: Classification of peat

1. Botanical composition • Moss peat (predominantly sphagnum and other mosses) • Sedge peat (sedges, grasses, herbs) • Wood peat (remains of trees and woody shrubs)
2. Degree of decomposition/humification (H) • Weakly decomposed (H1- H3) • Medium decomposed (H4- H6) • Strongly decomposed (H7- H10)
3. Trophic status • Oligotrophic (low in nutrients) • Mesotrophic • Eutrophic (high in nutrients)

Source: Kivinen& Pakarinen,1980

The type of plant remnants and their level of degradation affect the characteristics of peat. Peats vary in age and are frequently categorized using the straightforward but extremely useful von Post scale developed in the 1920s (von Post, 1922). This scale distinguishes between three types of peat: younger, undecomposed peat with low humification (H1-H3), partly decomposed peat (H4-H6), and older, highly decomposed peat (H7-H10). The test is making a fist out of the fingers while squeezing a small amount of peat into the palm of the hand. The degree of decomposition is determined by the color of the water expressed and the amount of peat that protrudes between the fingers. For horticultural use, peats with the designations H1 to H4 are sometimes referred to as "white peat," while those that are more decomposed (H4–H6) are referred to as "dark peat," and those that are extensively decomposed (H7 and above) are referred to as "black peat." In fact, attempts have been made to define peats more precisely based on measurements of color, and it has been shown (Prasad et al., 2015) that there is a strong correlation between color and humic acid content, providing the possibility of using a quick quantitative test to more accurately predict peat properties. Dark peats are less prone to biodegradation because they contain more lignin/humic concentration than white peats. Black peats frequently freeze throughout the winter, which increases their ability to hold both air and water. Earlier than extraction, peatlands are drained. For the purpose of extracting peat, machinery has been developed. Nowadays, peatlands that have been drained use milling technology to remove the top layer of peat, breaking up the semi-dried peat in the process. This layer, which is often just 1-2 cm thick, can be further dried before being gathered into big windrows or heaps. These piles are then sometimes covered with sheeting to avoid rewetting and kept in the field until needed. However, direct vacuum harvesting removes a significant amount of peat from drained peatlands. Since sods may be a superior source of big particle size fractions than milled peat, the practice of traditionally cutting peat into huge sods (usually 25 cm 3 10 cm 3 10 cm) that are subsequently air dried is still in use today. Due to its easy accessibility, consistency, and highly suited physical and chemical properties, peat has grown to be a significant part of growth media in Europe and North America. Great porosity, efficient aeration, and high water-holding capacity are all characteristics of moss peat. The addition of lime and fertilizers easily raises the low pH and low nutritional content to the correct levels. Peat is essentially devoid of weeds, pests, and plant infections, while it frequently promotes the rapid growth of the latter. Because of its modest weight, transportation costs are low. And finally, when employed as a growing medium, peat is comparatively stable.

The species composition of the bogs where peat is formed affects its physical characteristics because it affects the distribution of particle sizes and the level of decomposition. Peats are resistant to decomposition, especially degraded sphagnum kinds. Although sphagnum mosses lack the lignin found in vascular plants, lignin-like polymers make up a significant portion of their cell walls. However, the distinct pectin-like compounds (sphagnan) found in the cell walls of sphagnum mosses may potentially play a role in the resistance to breakdown. Additionally, phenolic antibacterial compounds are present in sphagnum peat. The bulk density is typically low and the total porosity is high, but the bulk density rises as the humification level rises. Due to its small weight, peat may be transported to nurseries and media producers both by road and ship at a reasonable cost. Since growing media based on peat can be well aerated and still hold an appropriate reservoir of water for plants, high porosity is the single most crucial physical characteristic. The physical makeup of the components as well as how they are extracted, processed, and prepared have an impact on the porosity of growth media. Sphagnum peat moss surfaces have holes that let water enter and flow through. Quantitative information on the size, geometry, connectivity, surface roughness, and anisotropy of the pore spaces as well as more in-depth knowledge of the porous structure of sphagnum peat in situ are now accessible. These data may be relevant to harvested peat utilized as a component of growing medium. In addition to relatively small, closed, dead-end pores created by plant cell remnants, peat also contains relatively big, interparticle pores that can actively transfer water. Smaller dead-end pores and somewhat large interparticle macropores can both be seen using scanning electron microscopy (SEM). Pore diameters in undecomposed peat may be greater than 5 mm, however with dewatering, compression, and decomposition, there is significant shrinkage. However, even after peat extraction and processing, the super-hydrophilic nature of peat is preserved, and peat media are able to absorb 10–20 times their own weight of water. Pore structure and size are related to the kind and degree of decomposition of peat. In its unaltered state, sphagnum peat with high fiber content and a lot of active pores can release up to 80% of its saturated water to drainage; heavily decomposed in situ peat samples can only release 10% of their water to drainage. After peat extraction and processing, this relationship is maintained: Younger, less decomposed peats (H2-H4) have a substantially higher water-holding capacity than older, more decomposed peats (H5-H7). The structural integrity of the latter allows air capacity to be maintained during usage of decomposed peats for long-term growth of large plants (shrubs, trees), as they may maintain higher air volumes than younger peats.

Numerous studies have examined the features of peat with varying humification values for horticultural usage, including bulk density, porosity, air-filled content, and water retention. When developing growing media, particle size is a crucial factor to take into account. For peat, this factor is influenced by the amount of extraction, processing, and decomposition. To provide peat growing media with a variety of physical qualities, peat produced by milling or cutting (to form sods) may be mechanically broken down and sieved to produce fractions of varying particle size. Either grinding the peat on the bog or creating sod peat and breaking these sods into the appropriate fractions can yield peat fractions. Bin sods and sod moss are the two different kinds of sods. Sod moss refers to sods cut directly from bogs of moderately decayed peat, whereas bin sods are formed by extruding peat with a high degree of decomposition (H6-H7) (H5-H6). Before usage, both kinds of sods are dried on the bog. Naturally, fractions from bin sods have a higher bulk density than fractions from sod moss. In comparison to fractions made from milled peat, those made from sods are more consistent and stable. The usage of fractioned peats can be used to increase air content inside media in big containers for the development of nursery stock, retain water, make plug trays easier to fill, and retain water in media meant for plant propagation. For plants that need more aeration, particularly those grown in big containers for months or even years, peat fractions are extremely useful. Consistent aeration during the growing season is crucial in these situations to guarantee ideal growth.

Growing media-grown plants must be able to digest the same variety of nutrients as soil-grown plants. For most plants other than calcifuges, the pH of peat-based media must be in the range of 5.5-6.5 for this purpose. The pH ranges for raw sphagnum peat (untreated, as harvested from a bog) are typically between 3.5 and 4.1. Desirable pH levels for the majority of growing media are easily achieved with peat through the addition of lime, commonly as dolomitic limestone (which also provides calcium and magnesium) at 2-3 kg/m3 for less decomposed (H2/H3) and 3-7 kg/m3 for more decomposed (H4-H6) peats. Sphagnum peat's acidic composition makes it a nearly essential part of media for calcifuge plant growth, including Azalea, Camellia, and Rhododendron species. Typically, peat-based media are dressed with 1-1.5 kg m3 of lime. In a few places, peats from fens, sedges, and reeds are utilized as parts of growing media. These peats can have a pH that is higher than sphagnum peat, up to 5.0 (in certain situations, high lime reed peats can be as high as 7.5). Although the cation exchange capacity (CEC) of growing media constituents is frequently reported on a weight basis, it is more appropriate (as with nutrient concentrations) to express CEC on a volume basis because plants grow in a finite volume of media and have low bulk density in the case of peat-based media. It is simple to convert from cmol (1)/kg (meq/100 g) to meq/L if the bulk density of the materials is known. The CEC of peat has been calculated to be 150–250 cmol (1)/kg for degraded H4–H6 peats, 80–130 cmol (1)/kg for sphagnum peat, and 80–80 cmol (1)/kg for sedge peat. These values roughly equal 150–250 meq/L and give many peats, especially H4–H6, an excellent buffering capacity, making them resistant to pH shifts caused by alkaline water supplies, and limiting the effects of fertigation on plant growth. Furthermore, peat has a good ability to retain cationic nutrients due to its high CEC values. The anion exchange capacity is, however, incredibly low. This implies, for example, that phosphate and N present in nitrate form can both be easily leached from peat-based growing media, unlike in most soils. Sphagnum peat's low pH and extremely low basic level of fertility necessitate the addition of fertilizers and lime in order to sustain healthy plant growth.

Of course, the species being cultivated and its stage of growth affect the amount of nutrients required. Nutrient requirements are much higher for plants that grow quickly than for seeds for propagation. However, base fertilization of media for seedling modules/plugs is typically low at 50–100 g L of a low-N fertilizer, but the exact rate used varies with different species. The cost and composition of proprietary fertilizers may dictate the type of these supplied to peat-based media. The base nutrient content of media for pot plants is higher, typically ranging from 1-1.5 kg/m3 of an 18:6:12 N:P:K fertilizer, although again, species might vary (Bunt, 1988). Similar levels of basic fertilizer may be present in the growing medium for large container plants, but this may also be enhanced with a suitable controlled release fertilizer. Additionally, or alternatively, plants in growth media may be fertigated using a suitable liquid feed. Peat-based growth media also require the addition of micronutrients. These can be included in compound fertilizers, individually as inorganic salts, or in slow-release forms, such as fritted trace elements (FTE).

Weed seeds from plants growing on peat bogs or the edges of bogs can contaminate peat. Rushes (*Juncus* spp.) and sheep's sorrel (*Rumex acetosella* L.) are the main plants in question. Control of these is often accomplished on-site through physical inspection and removal, with drainage ditches surrounding peat extraction areas receiving special attention. Peat develops in an acidic, wet environment that does not support widespread microbial growth. Total viable counts (TVCs) for harvested peat and peat-based media at the stage of manufacturing are typically 104 CFU g21 dry weight or less. While infections like *Pythium* or *Phytophthora* spp. can spread quickly in peat media once they arrive, the low population of microorganisms may not have the negative impacts that are seen in other media, particularly composted materials. Self-heating of stored peat, whether on the bog or in transit, is a significant issue despite a low intrinsic microbial population and is thought to be microbially mediated despite the low pH of peat. Thermophilic bacteria and fungus, including some uncommon species like the genus Alicyclobacillus, have both been linked to self-heating. Additionally, during self-heating, phytotoxic chemicals may be released, self-heated peats' physical, chemical, and microbiological properties may change, and plant germination and growth in media made from such peats may be negatively impacted. In fact, self-heating peats are typically unsuitable for use in growth medium manufacturing. Structural stability is necessary for the preservation of physical qualities, particularly air space, inside the medium during plant growth. The structural stability of any organic-based growth media may be negatively impacted by microbial deterioration, along with shrinkage and swelling. Peat media degrades more slowly than media made of other ingredients due to the presence of microbes. Oxygen uptake rate (OUR) measurements of biological stability show O2 consumption rates as low as 1.9 mmol/kg/h. Irish peat (10-30 mm) showed the lowest rate of respiration at 16.1 6 3.1 mg CO2-C/kg/day, followed by sod peat (10-30 mm) at 25.2 6 1.5 mg CO2-C/kg/day and coconut fiber at 82.6 6 4.0 mg CO2-C/kg/day, according to research by Grunert et al. (2016). The quantity of lignin-like compounds found in various peats has been connected to this slow pace of decomposition. Sphagnum peat also includes humic chemicals that are extremely resistant to microbial destruction and phenolic antibacterial compounds. As a result, moss-derived peat shrinks very little when used, even when the pH is elevated and the media is fertilized. In comparison to alternative substrates, humified sphagnum peat is frequently favored for growing nursery stock plants in big pots or containers over months or even years.

Coir

The mesocarp, or middle layers, of coconut fruits are made of coir (Cocos nucifera L.). These layers are made up of fibers that are enmeshed in the so-called coir pith, which is also known as coir dust when it is dry. One of the most prevalent organic wastes made by plants is coir, which is found in many tropical and subtropical nations. After soaking, long fibers are removed from the mesocarp and used to make mats, brushes, and insulating materials. The remaining material is made up of coir pith and short fibers. This material was historically abandoned and accumulated as waste, and it is still seen as an issue in some places. However, these coir "dumps" have almost disappeared from the main areas of production, and the fabrication of substrates now primarily uses coir pith from recently harvested coconuts. Fresh coir is often matured or "composted" for several months after fiber separation before being prepared as a substrate for growing media. Coir wasn't widely used as a growth media until the 1980s, partly because of issues with quality and consistency, despite being investigated as a potential growing medium in the 1860s with interest reawakened in the 1940s. Improvements in quality assurance have led to higher output levels of coir, which is used as growing material. Prior to being prepared for horticultural use, coir pith is washed with water and/or solutions, such as Ca(NO3)2 (the so-called buffered coir), because it may contain high salt levels (mostly Na, Cl, and K). The dried coir pith is crushed (varying from 5:1 to 8:1 v/v, Ross et al., 2012) into blocks (about 5 kg) or briquettes (approximately 0.6 kg) for convenience of transport after washing or leaching. Blocks are reconstituted with water after being received by expanding media producers, with manufacturers aiming for an expansion ratio of 1:12 (wt: vol) or higher. Additionally, coir chips (5–15 mm) made from the coconut's tough outer husk are employed in growing medium, mostly to improve aeration, especially when it comes to media for long-term nursery crops. The Tamil Nadu and Kerala regions of India, as well as Sri Lanka, are important producing regions; coir is predominantly exported from these regions to Europe and Australia. Coir is a product that Mexico sends to the US. Coir pith exports from Sri Lanka increased from 97,000 tons in 2007 to 120,000 tons in 2010 (equal to 1.16 M m3 if reconstituted on a 1:12 wt: vol ratio) (Anon, 2011). From 6501 tons in 2000-01 to 408,897 tons in 2015-16, India's exports of coir pith, which are mostly used in horticulture, have expanded significantly. If reconstituted at a ratio of 1:12 (wt: vol), this corresponds to volume increases from 78,000 m3 to 4.8 M m3 (Fig. 8.3). In 2015–16, the Netherlands alone imported 73,735 tons of coir from India, or over 885,000 m3 when reconstituted at 1:12 (wt: vol). The Philippines, Indonesia, Ivory Coast, Costa Rica, and Thailand are some other nations growing coir for horticulture usage. Brazil produces a lot of coir, although most of it is used as a soil amendment.

Coir has been a popular choice for organic substrate due to improvements in supply and quality control. Coir has a very high water-holding capacity, great porosity, and effective aeration. Similar to peat, coir pith is virtually free of weeds, bugs, and plant pathogens. To verify their goods are free of harmful organisms, the majority of large exporters work with organizations like the Netherlands-based RHP (Regeling Handels Potgronden). Finally, coir is a relatively durable growing medium that can even be used again in some horticultural applications, such as growing strawberry plants in modules. Transport expenses for growth media producers in North America and Europe are undoubtedly higher than for peat. Due to source, age, and degree of processing, sources of coir can vary substantially in physicochemical (pH, electrical conductivity - EC) as well as chemical and physical characteristics. It is crucial that authors note the latter when reporting tests of coir, however, this is not always done.

The composition of macromolecules and how they are arranged within the coir pith may help to explain why it is such a good substrate for plant growth. Depending on the age and treatment (especially composting) of the coir, several major elements are present. The main macromolecular component of coir pith is lignin, which is primarily p-hydroxyphenyl-guaiacyl-syringyl lignin in the coir fiber. The weight-based lignin percentage of coir pith has been estimated to range between 30 and 50 percent, and the lignin structure can deteriorate over time, especially in heaps used for storage or composting. The amount of cellulose in coir pith varies from 26 to 35 percent by weight according to different writers, with some citing levels as high as 50 percent. At 15 percent to 35 percent, hemicelluloses make up the majority of the residual polymers in coir pith. It is unclear how these macromolecules are arranged in coir pith, but they must be connected to the features of pith that make it such a desirable choice as a substrate for plant growth. The lignocellulose complexes, made of cellulose, hemicellulose, and lignin connected through covalent and noncovalent connections and highly resistant to degradation, may be linked to the resistance to coir pith degradation. Also possible in coir are phenolic compounds. Coir pith is extremely hydrophilic, just like moist peat is. It has been demonstrated that coir from various sources may hold 5–10 times its weight in water. Due primarily to their microporous nature, which permits internal retention of both water and air, peat and coir are both well known for their capacity to absorb and retain both water and air. Few coir pith ultrastructural studies of pore properties have been conducted. Coir dusts from Mexico and Sri Lanka did not differ in microstructural features, although they did differ significantly in particle size, according to a study by (Fornes et al 2003) employing SEM. According to Fornes et al (2003), the external circular pores in coir pith have a diameter of 30 to 80 m and produce a relative surface porosity of about 49%, which is significantly higher than that of sphagnum peat, which has a 12% porosity. The more frequent and (bigger) size of holes on the surface of coir particles in comparison to sphagnum peat particles may account for the faster uptake of water, similarly quick drainage, and consequently higher air content than other peats.

Contrary to sphagnum peat, coir is rarely hydrophobic; hence, wetting agents are not required in media that include mostly coir, but water-absorbing polymers are routinely added to peat-based media to aid in absorption after drying. The actual husks from which coir pith is obtained, processing of these, in particular the level of separation of pith and coir fiber, as well as the age of the pith, all contribute to this variance in the overall physical qualities of air and water capacity of coir pith. The physical characteristics of coir may be significantly influenced by its particle size. According to Evans et al. (1996), samples of coir from the Philippines primarily had pith particles between 0 and 2 mm in diameter, with one source having a high concentration of very fine particles between 0.25 and 0.5 mm in diameter. This source also had the highest bulk density, water-holding capacity, and total pore space of the five samples analyzed, but the lowest air porosity. Abad et al. (2005) discovered that most samples of coir from various nations had particle sizes that ranged from 0.5 to 2 mm, although there were several notable outliers. Abad et al. (2005) only looked at one sample of Indian coir, while Ross et al. (2010) discovered that five samples of coir pith from Tamil Nadu mostly included particles with a size between 0 and 3.2 mm.

The large difference in airspace and readily accessible water in coir may be due to particle size, which is significantly correlated with aeration and water availability, especially with particles of less than one millimeter. When specific specifications are needed for specific growth systems, the particle size of coir can be modified by sieving at the site of production, according to the standard protocols of many coir exporters, which generally describe a particle size of primarily 0.5-1 mm.

Generally speaking, coir is less acidic than peat, though this can vary greatly depending on the source. Lime may not be required if coir makes up the majority of a medium, but calcium must be supplied, and gypsum is frequently suggested for this purpose. Depending on factors including age, processing, if and how often it was washed, and/or buffered, the EC of coir can vary greatly. Comparing published EC data is challenging due to the diversity of methods used globally to quantify the EC of growth medium and their constituents. EC values for raw coir samples from Tamil Nadu, India's east and west coastlines, were as high as 11.25 dS/cm and 8.08 dS/m (1:1.5 extracts), respectively. Sequential washing of coir significantly lowers EC values; for instance, in a sample of Indian coir with a particle size range of 0.2 to 0.5 mm, after five sequential washes, the EC value dropped from 6.5 to 0.14 dS/m (1:1.5 extracts) while the pH increased from 5.9 to 7.8. Natural high concentrations of Na, Cl, and K in water-soluble forms are the main cause of elevated levels of EC, and it is the leaching of these elements that causes the decrease in EC. With values ranging from 35 to 95 cmol (1)/kg (about 70150 meq/L), coir has a lower CEC than peat. When it comes to macronutrients, coir sources are typically low in N, with most instances recording both NH4N and NO3N at.5 mg/L. Some coir sources have considerable P concentrations up to 30 mg/L, while the majority have far lower amounts. Nearly all samples of unprocessed coir are high in K, Na, and Cl, and as mentioned, these cations need to be removed before the material is ready to be used as a substrate for horticulture. When washing water contains Ca(NO3)2, exchangeable Na and K are eliminated, and the exchange complex is primarily saturated with Ca. Moderate-to-low residual concentrations of these ions may persist in coir substrates even after washing, particularly with water, and in fact, the majority of importers of coir pith for horticultural uses will specify cation concentrations and EC limits to exporters. Micronutrients may also be present in coir, again depending on the source, but their concentrations might vary greatly. Despite some documented anomalies, processed and exported coir pith often does not include weeds or bugs. In contrast, raw coir has a microbiological content that is at least two orders of magnitude larger than raw peat, with TVCs of about 105 (CFU) bacteria, 104 actinomycetes, and 103 fungi per gram of raw coir.

Although coir, like peat, is regarded as a stable substrate component, microbial activity in coir-based media has been found to be slightly higher than in peat media. Studies on the potential of coir for the biocontrol of plant pathogens have been prompted by the greater microbial populations in coir than in peat. Even though this research was only done in vitro, it has been demonstrated that some wilt fungi can't develop in the presence of coir extracts. It has been shown that adding biocontrol agents successfully controls disease in growing media containing coir (Castano et al., 2013). Around the world, coir is widely used as a plant substrate. For many crops, it is included as one of the growing media's components, such as in nations in the Southern Hemisphere along with bark or in Europe as a peat diluent. The water-holding capacity and wettability of coir make it an appealing option for bedding and container plants, whose development has frequently been proven to be superior to that in peat. Coir is highly prized as a rooting medium. Growing soft fruit crops like strawberries, raspberries, and blueberries as well as roses, lilies, and some other ornamentals on coir has become more popular in the 2000s. Systems like the so-called table-top culture of strawberries grown in bags or premade modules containing coir have replaced soil-based systems, where pest and disease problems, in particular, have become harder to manage as a result of restrictions on pesticide use. Because coir can be used three or four times in growing modules for soft fruit crops, the increased cost is offset.

Bark

Bark, a by-product of sawmills that is frequently referred to as "residue," comes from both softwood (mostly conifer) and hardwood (mainly in North American) species and is a significant portion of growing media, particularly in regions where peat is either expensive or unavailable. Prior to the 1960s, bark in the United States was discarded as a byproduct of the lumber industry. However, as its usefulness as a decorative mulch and element of growing media came to light, the use of pine bark in particular progressively rose. Nursery stock species are frequently produced in media where bark is the main component in the US, Australia, and New Zealand. They are raised in enormous containers over several months or even years. Bark and peat are frequently combined in Europe for the cultivation of several species in big containers. Raviv et al. published one of the earliest reviews of bark as a growth medium (1986). Although bark that has been aged can be used in place of fresh bark from sawmills as a primary ingredient in soilless substrates used for container nursery development. Material that has been piled up for several months or perhaps years is referred to as aged bark. Potentially hazardous substances like phenols and terpenes may disperse during this time, and low-molecular-weight organic substances may be ingested by microorganisms, resulting in biological stability. Bark that includes aging is particularly appropriate. 10% so-called "white wood" (xylem): This bark, which has a high cellulose/hemicellulose composition, has the potential to immobilize nitrogen extremely quickly when applied fresh. For the production of composted bark, which is mostly utilized in Europe and Australia, windrow preparation, nitrogen application, turning, and maturation are required.

Bark that has aged and especially composted is significantly less hydrophobic than new bark. However, regardless of treatment, bark will turn hydrophobic if dried to 35 percent (w/w) or less. When composting bark, more nitrogen is added to help overcome N immobilization and lower toxicity. Other substances are normally not included while composting, though phosphorus is occasionally advised as well. Moisture, gaseous oxygen, and temperature are other crucial factors in composting. The ideal rate of urea addition was thought to be 3% of the initial bark dry weight. Composting reduces the amount of organic molecules that decompose quickly. From 2005 to 2013, the amount of bark used in growth media in Europe stayed at around 1.9 M m3. France, with 1.05 M m3 in 2013 (down from 1.2 M m3 in 2005), is the main consumer of bark in Europe. Bark used in France is primarily obtained from the maritime pine, *Pinus pinaster* Aiton. In addition to *Pinus pinaster* and *Pinus elliotii* Engelm, hardwood bark from *Eucalyptus diversicolor* F. Mueli and *E. (Corymbia) calophylla* Lindley is also used in Australia. In New Zealand, *Pinus radiata* D. Don's Monterey pine bark serves as the main organic component of growing media. The main sources of bark for use in growth media in the United Regions are the Douglas fir, *Pseudotsuga menziesii* (Mirb.) Franco in the Pacific Northwest and the slash pine, *P. elliotii*, and loblolly pine, *Pinus taeda* L. in Eastern states. The northern United States and Canada use other coniferous species such the red pine, *Pinus resinosa* Sol. Ex Aiton, and the jack pine, *Pinus banksiana* Lamb. Hardwood bark from redwood (*Sequoia* spp.) and other species is also used as growing media in several regions of the United States. Due to a fall in sawmill activity after 2008 and a rise in the utilization of wood, including sawmill residues, for pellets used as biomass, bark supplies for growth media, particularly in the Southeast United States, decreased. But in the second decade of the new century, bark continued to be the most often used substrate in nursery stock media in the US. Due to debarking occurring in the forest rather than at sawmills, bark supplies have also decreased in France, raising the cost of bark recovery. In Australia, issues with bark supply for growing media have also been brought up.

Bark varies greatly in terms of morphology and constitution depending on where it comes from. When *Pinus* spp. bark is examined under light and electron microscopes, it is discovered to have a porous structure that is connected to the voids within dead cells, primarily phloem, axial, and radial parenchyma. According to Pokorny and Wetzstein (1984), the internal porosity of *P. taeda* bark accounts for 42–44% of its volume, and plant roots have the ability to enter these pores and absorb water there. In fact, since bark substrates are often free-draining and have a lesser water-holding capacity than either peat or coir, water kept internally inside the bark may provide a significant reserve. Peat can be added to bark mixes to increase their water-holding capacity, which is typically low and has a high percentage of air-filled porosity. Bark is ground or hammer-milled, and then screened to generate a variety of particle sizes. This process enables the production of specified combinations of air and water capacity. Pine bark is typically screened between 0.5 and 10 mm for nursery stock in the United States. In Europe, spruce bark may be screened to a thickness of 10–20 mm for nursery stock and 2–10 mm for general propagation and/or container plants. Although aged bark has less air than young bark, it has a much greater capacity to hold water. On a dry bulk density basis, bark from most sources is heavier than sphagnum peat or coir. According to the species, growing conditions, age of the trees, the time of harvest, and the methods of harvesting, particularly the effectiveness of separating bark from underlying wood, bark's chemical makeup varies. Like peat and coir, lignin is a key component of many polymers. Suberin and other lignin-like substances are often included in lignin studies for bark. In *Pinus* spp., lignin and related polymers are so frequently reported to be as high as 40–50% w/w. In addition to cellulose, hemicelluloses and, in certain circumstances, pectins are also present in the bark of Norway spruce *Picea abies* (L.) H. Karst, *P. taeda*, and *P. menziesii* species, according to monosaccharide studies of these polysaccharides. Additionally, mono- and disaccharides, which are present in relatively high concentrations in bark, may be absorbed by microbes, which may also be responsible for the cellulose's breakdown, resulting in nitrogen immobilization. Although there is little information on the breakdown of hemicelluloses in bark, these substances are easily hydrolyzed in other systems, and their breakdown after microbial colonization of growth media containing bark may also help to immobilize nitrogen. Additionally, barks contain phenols, condensed tannins, and other substances that are typically categorized as soluble and/or solvent extractable materials. Some of these substances are extracted and used commercially, but at least in fresh bark, they may have significant phytotoxic effects that affect plant growth and germination. Such effects will be lessened or eliminated by aging and/or composting. A variety of inorganic chemicals are present in barks, with some species having higher quantities than others (Lambert, 1981). Macro-elements (P and K) are present in barks from a variety of sources, as shown by EC values. K concentrations are such that base fertilization with this element could be reduced. Although better development has been seen at pH 5 where minerals have been added to pine bark medium, barks may also contain enough micronutrients to meet the needs of many plants. In fact, some micronutrients may be present at levels that could prevent plants from growing. For example, manganese levels in spruce (P. abies) bark have been demonstrated to prevent plants from growing; in these situations, liming is advised to precipitate the manganese. Lime may be given at lower rates than peat since the pH of most barks—4.0–5.0 for pine bark and 5.5 for spruce bark—is higher than that of peat. In certain situations, plant development in pine bark media at pH 5 did not respond to liming, and in fact, in some species, growth was inhibited. In general, aged bark has a somewhat lower pH than fresh bark, although composted bark's pH might be as high as 5 or 6. The CEC of barks is often lower than that of sphagnum peat.

It has been widely documented that nitrogen immobilization occurs in bark-based media and is related to the colonization of the bark by microorganisms that can assimilate free mono- and disaccharides as well as degrade polysaccharide components like cellulose and hemicelluloses, assimilate the monosaccharides hydrolyzed from these polymers, and colonize the bark. The acquisition of nitrogen as well as other elements like phosphorus, potassium, and trace elements is necessary for the growth of microorganisms. In bark media, phosphorus depletion is possible (Handreck and Black, 2010), but nitrogen immobilization is more frequent, especially in substrates made from fresh bark. It is possible for nitrogen immobilization to continue even after composting, for instance with spruce bark. Bark is typically treated with urea at 2-4 kg m3, depending on the species, when composted (typically for 4–8 weeks). N can be supplemented to media even after composting. Nitrogen immobilization differs between species, with spruce bark being more susceptible to depletion than pine bark, possibly due to the much larger content of free sugars in spruce bark. However, growth reductions in substrates with added bark could not always be the result of nitrogen shortages and, in certain cases, might also be the result of insufficient root aeration or even overwatering. The antagonistic actions of microorganisms present, particularly in composted medium, are thought to be responsible for barks' capacity to limit the growth of plant diseases in growing media. As a result, root diseases including Pythium and Phytophthora damping off and root rots, as well as Fusarium wilts, can be suppressed using barks from pine and hardwood trees. Good drainage helps to lessen sporulation and the transmission of waterborne Oomycete pathogens in media, in addition to directly suppressing disease and fostering root development encouraged by high air content.

As previously mentioned, bark is a significant component of nursery stock media in the United States, Australia, and Europe. It is also used as a container media component. In the UK and Ireland, spruce bark is a significant part of peat-free media intended for retail and hobby markets, whereas P. pinaster bark is extensively utilized in France for both professional and hobby markets. Bark works well as a propagation medium in both high humidity and intermittent mist situations. In order to create nursery stock media, bark is frequently combined with other organic as well as inorganic ingredients. Sand (0.5-2 mm) is a popular choice in the United States because it helps to stabilize containers and hold in moisture. Peat and other materials like cotton waste and whole tree substrates are also frequently used. In container medium for shrubs and trees, bark is frequently combined with peat and grit (2–5 mm) in Europe. For epiphytes like orchids and bromeliads, bark can be used in specialized ways, such as in the form of bark chips (20–40 mm). Typically, steam is used to disinfect chips for orchids. In North America, Douglas fir (*P. menziesii*) bark chips are frequently used for orchids; in Europe, *P. pinaster* is the main species. For the growth of genera like Cymbidium, Douglas fir bark is typically graded to 6-7 mm; for Cattleya, Phalaenopsis, and Dendrobium, to 12-15 mm; and for Vandas, to 20-25 mm. The high air content required for orchid growth is provided by growing media made of bark chips, but the chips decompose slowly—possibly because antimicrobial compounds are present—so they are not composted. In an effort to reduce water loss during transit of these highly valuable ornamentals, attempts have been made to include coir and other water-retentive substrates into the medium for orchid culture. Orchids raised in high particle size bark chips require frequent irrigation. The growth of orchids like Phalaenopsis has been made possible by materials like undecomposed sphagnum moss and wood fiber mixed with pine bark. These materials are equivalent to or even superior (in terms of root growth) to traditional combinations of pine bark and sphagnum peat (70:30 v/v).

Wood-Derived Materials

Wood fiber is produced from clean, chipped wood while sawdust is a byproduct of the timber industry. Usually, full trees as well as leaves, twigs, branches, and needles removed during forest thinning are ground into powder to create whole tree substrates. In several nations, especially those where peat is in short supply or expensive, sawdust has been utilized as a component of media, including Australia, New Zealand, the United States, South Africa, and Argentina. Sawdust is made from a variety of species, including conifers like *Pinus radiata* in Europe and Australia, *P. menziesii* and *Tsuga heterophylla* (Raf.) Sarg. in the United States, and *P. radiata* in Australia, New Zealand, and South Africa. Additionally, hardwood species' sawdust has been included in growing media. Before being added to planting media, sawdust may be composted or weathered outdoors for a year or longer. Mechanical defibrillation, or more frequently steam-assisted thermal extrusion of clean wood chips using a thermo-screw press, is used to create wood fiber. The product is heated to 80°–90°C during processing due to friction, making it free of infections and bugs. The procedure may also drive off volatile poisons. The device can additionally infuse wood fibers with chemicals like nitrogen fertilizers. Additionally, coarse or fine grades of wood fiber can be generated. Wood fiber can be compressed for easier transport, just like coir. The first wood fiber products were Horti-fibers from France and Toresas from Germany. Fibralurs, a different type of wood fiber, was created in Spain in the 2000s. Between 2010 and 2017, there was a significant increase in the production and consumption of wood fiber, with new manufacturing facilities opening in the United States, Ireland, Germany, and Latvia. Wood fiber volumes used in growth media increased in the United Kingdom from 62,000 m3 in 2007 to 0.5 M m3 in 2013. The early 2000s saw a rise in interest in the use of whole tree substrates, particularly pine tree substrates and related clean-chip residual materials. This interest was further stoked by the decline in sawmill activity from 2005 to 2008, which may have resulted in a decrease in the supply of bark for horticultural use. Pine tree substrates are made from either clean-chip residuals or pine trees that have been chipped and processed in a hammer mill (with or without bark, limbs, or needles) (approx. 40 percent pinewood, 50 percent bark, and 10 percent needles).

Despite having a limited water-holding capacity and good aeration, wood-based materials vary greatly in their particle size and shape. Sawdust has the benefit of having a low bulk density and is frequently made up of tiny particles with mean values normally ranging from 1 to 5 mm, while fractions sieved to 6 mm have been employed for crops like strawberries. Water stress during vigorous plant growth may occur from the low available water content, hence irrigation must be administered often in modest quantities. Low bulk density, high total porosity, and extremely high air content are the characteristics of wood fibers used in growing media, which results in a higher oxygen diffusion rate than peat. Processing influences physical attributes to some extent. Wood fiber is typically not utilized as a stand-alone growth media component, with the exception of bagged modules for glasshouse vegetables, because it holds insufficient plant-available water and has a propensity to become compressed while being employed. Similar to bark, the degree of processing and, in particular, the size of the sieve used to produce the final product determine the physical qualities of entire tree substrates. Depending on the source, reports range from neutral readings of 6.3 to 7.7 to higher acidic values of 5.33 and 4.71 for sawdust. The pH ranges for wood fiber are 3.8 to 6.6. Sawdust and wood fiber don't have much buffering power, thus the materials they are mixed with or the pH of the liquid fertilizer they receive during plant growth may have a greater impact on their pH. The CEC of sawdust, wood fiber, and entire tree substrates is in fact low; the relatively high figure for wood fiber is likely the result of sample grinding to a thickness of 2 mm. With the exception of K in some materials, macro and micronutrient concentrations in water or CaC12/DTPA (CAT) extracts are typically quite low for all wood-based substrates.

Similar to several barks, uncomposted sawdust significantly restricts plant development. Plant development can sometimes be inhibited by phytotoxic chemicals, most notably when cedar sawdust is involved, however nitrogen shortage is typically to blame. A significant issue with wood-based substrates is the immobilization of nitrogen supplied to plants in the substrate solution, along with media degradation and shrinkage. This issue may be related to the proportions and possibly relationships between the polysaccharide and lignin content of these materials. Wood-based materials have lower lignin contents than peat, coir, and bark, and one study (Domeno et al., 2010) found that hemicelluloses are present in very high concentrations. In wood-based substrates, particularly those with small particle sizes like sawdust, fine wood fibers, and finely crushed whole tree substrates that have a lot of surface area for microbial colonization, the later polysaccharides may be easily destroyed. Similar to bark, microorganisms that colonize wood substrates get their carbon primarily from the polysaccharide elements and their nitrogen from fertilizers added to growing media. Nitrogen drawdown, which occurs when substrate components made of wood or other organic matter lose nitrogen, calls for additional N in the form of liquid feed or controlled release fertilizers. Decomposition of sawdust-containing medium has been found to cause volume losses as high as 50% over a year. Additionally, decomposition results in slumping and a consequent drop in air content.

Therefore, nitrogen fertilizers must be used at high rates to successfully produce plants, especially those grown on sawdust made from Eucalyptus species. Because of their high price, even composted sawdust may have a rate of nitrogen immobilization that prevents them from being used as media for bedding plants and plant propagation (Handreck and Black, 2010). According to Prasad (1997b), nitrogen immobilization brought on by microbial activity was quite evident in a number of wood fiber-based growing mediums, with accessible N levels dropping by as much as 150 mg/L over the course of 5 weeks of incubation. With urea, N release was significantly faster than with hoof and horn fertilizer, the latter evidently being destroyed more slowly by the indigenous microbial flora of the wood fiber-based medium. The release of N from organic sources of N in wood fiber-based media varied. Shrinkage results from the decomposition of wood fiber-containing media, and it is much greater than in peat-containing media. The immobilization of nitrogen in wood fiber can be somewhat mitigated by adding nitrogen during the production process. Untreated wood fiber quickly immobilizes nitrogen, but N-treated wood fiber immobilizes nitrogen more slowly, more like H2/H3 (white) peat. The growth of tomato transplants grown on nitrogen-impregnated wood fiber was noticeably higher than that of plants grown on unimpregnated wood fiber. However, Gruda et al. (2000) advise fertilizing plants grown in wood fiber substrates from the time they are planted with nutrient solutions or fertilizer. Clean-chip residual wood substrates may immobilize nitrogen, according to Boyer et al. (2012). Additionally, they discovered that immobilization was much higher than in pine bark and that microbial respiration in clean-chip residual was much higher than in peat. Additionally, respiration was more prominent in substrates with clean-chip residuals and smaller particle sizes. Therefore, more N is advised for these items. Despite N immobilization, after 2 years in huge containers with woody nursery crops, there is no discernible degradation or shrinking.

In bag or module culture, strawberries and greenhouse vegetables have both been grown using sawdust, either alone or in mixes. Wood fiber has been utilized successfully with a variety of protected food crops, similar to how sawdust has. Plants in substrates containing.50% of wood fiber had to be watered more frequently than those in substrates containing peat because the amount of water available decreased with increasing wood fiber content, but overall, the total irrigation water required per growing season for wood fiber was lower than that required for peat. To generate growth media for retail markets in the UK and Ireland as well as for usage as a component of media for bedding plants, wood fiber is frequently combined with peat. With entire pine tree/clean-chip residual substrates, a variety of greenhouse crops, including chrysanthemums, poinsettia, annual bedding plants (including ageratum, salvia, and impatiens), vegetables, and herbaceous perennials, have been successfully grown. In this media, nursery stock and tree species have been successfully raised. The root development in substrates made from whole pine trees in the case of annual and woody plants is said to be comparable to or superior to that in peat or pine bark. However, immature marigold and tomato seedlings in plug trays occasionally showed growth loss when freshly harvested trees were used to generate substrates; as a result, the use of aged whole tree substrates for young transplants was advised.

Rice Hulls

The outermost layer of the rice grain (*Oryza sativa* L.) is known as the hull, which serves as a protective covering for the grain. After parboiling, the rice grains are separated from the rice hulls by milling (soaking, steaming, and drying the grain). 770 Mt of rice hulls are produced annually in Asia, where rice production, and are available in huge quantities as waste from rice processing industries worldwide. A large portion of this is burned as fuel, and the ash is utilized to make building materials. As much as 1.9 Mt of rice hulls may be produced in the US each year as a result of rice processing, and some companies here use these hulls as part of their growing media. In 2013, professional growers in the Netherlands used about 12,000 m3 of rice hulls. It has been thought of using aged, composted, or burnt (carbonized) rice hulls as a growth medium. It has been proposed to use biochar made from rice hulls as a growth medium. With typical dimensions of 34 by 57 mm and dry bulk densities of uncompressed, unground material ranging from 100 to 400 g/L depending on the degree of milling, grinding, and sieving, rice hulls have an elongated convex form. In addition to cellulose (30–40% by weight), hemicelluloses (15–25% by weight), and lignin (15–25% by weight), rice hulls also include around 15% by weight of silica. These last two elements give rice hulls a high degree of durability and gradual degradation. Fresh rice hulls have a pH of 4.64.8, which increases to 6.57.0 when they are stored. The P concentration may have been significantly lower as a result of the water extraction method used. Rarely are rice hulls utilized as the only ingredient in growth media.

Composted Substrates

Before using them as growing media, materials like coir, rice hulls, and especially bark, can be composted. The primary global goals of composting include handling organic wastes and byproducts. Volume reductions, the removal of harmful organic compounds, and the use of readily available carbon sources may provide final products that can be used as growth media. Many nations have access to substantial volumes of composted materials, and a variety of composted wastes and by-products have been looked into for use as growing media. However, composts are the most variable of all potential growing media in terms of bulk density, airspace, and water retention; chemically in terms of pH and nutrition; and also microbiologically (Raviv, 2011). Compostable materials are almost always utilized in conjunction with other materials (bark is the exception). In the Netherlands, United Kingdom, Italy, and Germany, composts made from so-called "green materials" (prunings, shredded branches, plant debris, and trash from gardens and nurseries) are frequently used as growing media components, mainly in media for the hobby market. Germany saw an increase in compost use in growth media from 0.25 M m3 in 2005 to 0.7 M m3 in 2013, the Netherlands had an increase from 10,000 m3 in 2001 to 95,000 m3 in 2013, and Italy saw a rise from 0.29 M m3 to above 0.29 M3 in 2013. In 2013, 0.31 million m3 of compost was used in the UK as growth media. In some southern European nations, such as Spain (Moral et al., 2013), as well as in Canada and the United States, where composts are frequently employed as components of media for the development of nursery stock, municipal wastes and sewage sludges are included into growing media. Strict quality control procedures are necessary when preparing composts for use in growing media. The Waste and Resources Action Program (WRAP) in the United Kingdom and the RHP foundation, based in the Netherlands, which is the main organization in charge of overseeing growing media standards in Europe, both set acceptable parameters for composts destined for use in growing media.

The composting process is the biological breakdown of organic materials into a stable humus-like product under controlled conditions. A thorough composting process that includes a long thermophilic phase can minimize phytotoxicity, eliminate pests, pathogens, and weed seeds, and stabilize the final product's nitrogen and oxygen requirement. Composting depends on the C-to-N ratio, and the ideal range for the source material is between 25 and 35. The pace of decomposition will be extremely sluggish and the resulting compost may not be stable if the value is far higher than this. Ammonia (NH_3), which is a type of nitrogen, is frequently lost to the air with lower C: N ratios. By assessing the feedstock and adjusting its C or N levels with the addition of C-rich materials like wood or pruning waste or N-rich substances like poultry manure, the ideal C: N ratio can be attained. It may also be important to consider the type of C and N present, as lignin-rich materials and nitrogen found in protein-based waste from municipal solid wastes, cheese whey, and other dairy activities may decompose slowly in composting systems. As CO_2 is released during the composting process, the C: N ratio decreases, and a final value in the range of 15-20 is often desired. To speed up the composting process, the materials to be composted are first shred in order to minimize particle size and encourage microbial penetration into the substrate. Additionally, shredding makes handling the material easier and makes prescreening and combining easier. The temperature will start to climb as microbial activity increases after shredding, adding any materials like N and water, and often shaping into compost mounds. The thermophilic stage, which lasts until the easily decomposable organic matter has been broken down, normally begins after a few days and reaches temperatures of over 60°C. The material must all be exposed to temperatures that are close to one another for uniform composting to occur. This is achieved in many systems by stirring the compost pile on a regular basis to make sure there is enough oxygen available for aerobic degradative processes to take place. Temperature homogeneity throughout the pile is a crucial factor in the design of the composting system where turning is not done, such as in static piles. Temperatures should not rise over 60 to 65 °C because ashing or loss of organic content may occur. This should be prevented by turning and/or forced aeration. Once the easily decomposable material has been broken down by microbes, the temperature will start to drop, and as soon as it drops below 40 °C, mesophilic bacteria will re-invade the compost, reducing ammonia loss from nitrification of ammonium ions. The active composting phase is replaced in the final step by the curing phase, during which the compost reaches maturity and is prepared for use as a component of a growing medium. This will take 12–20 weeks for the majority of materials in regularly turned stacks. Compost exhibits suppressiveness against a range of soil-borne diseases due to some of the microorganisms that infect it during the curing phase.

When using compost as a part of a growing medium, maturity and stability are crucial qualities. Active decomposition can negatively impact plant growth if it goes on after being incorporated into a growing medium since it will result in lower oxygen and nitrogen availability as well as the presence of phytotoxic chemicals. In order to assess the stability, maturity, and phytotoxicity of composts, tests have been created. Respire-metric measurements based on O2 intake and CO2 evolution are frequently used to assess stability and maturity. It has been proposed that a particularly sensitive indicator of compost maturity is the rate of heat production using microcalorimetry. Different nations utilize chemical tests to determine the compost's maturity, including measurements of the nitrate/ammonium ratio, C: N ratio, and volatile organic acid content. A bioassay is typically used in phytotoxicity assays to check for adverse effects on seed germination and plant growth.

The physical and chemical characteristics of composted materials are often determined by the nature of the source materials and how they were processed. Pore space is high in materials containing animal dung, such as spent mushroom compost (SMC) and many green composts, but not all of them. Air content varies as well; it is higher in materials made from grape marc and olive waste, but water content is lower. Compared to peat or coir, the majority of materials exhibited residual lignin values between 15 and 25 percent. The bulk densities of all the materials were significantly higher than those of ordinary peats, with the exception of a few SMCs, which is a common issue with most composted materials. All of the materials had rather high ionic concentrations, particularly K, and high EC values, which basically prohibited their usage as individual media components and really limited their levels in growing media. The pH levels were also generally high. Some composts may have nutritional ions (NH4, NO3, P, K) and micronutrients that allow for lower fertilization rates in growing media, however, composted materials can also have unacceptable levels of Na and Cl as well as high levels of heavy metals. In this regard, the Netherlands-based RHP and the WRAP in the United Kingdom have established acceptable ions concentrations for composted materials intended for use in growth media. Pathogens from plants, people, or animals may contaminate waste materials that are intended for composting, raising concerns that these organisms might endure the composting process. Jones and Martin (2003) came to the conclusion that the majority of pathogens are rendered inactive by a composting process with a residence period of 3 days at a temperature greater than 55°C after reviewing the survival of human and animal pathogens in composted green waste (CGW). In their reviews of 60 plant disease and nematode species, Noble and Roberts (2003, 2004) examined the temperature/time effects and other sanitizing aspects of composting. A compost temperature of 55°C for 21 days was adequate to achieve the eradication of all bacterial plant pathogens and nematodes, the majority of fungal plant pathogens, and a number of plant viruses. Some diseases with extremely resilient spore and sclerotia resting bodies, however, proved more challenging to get rid of. *Plasmodiophora brassicae* (clubroot of Brassica species), *Fusarium oxysporum f. sp. lycopersici* (tomato wilt), and *Macrophomina phaseolina* (dry root rot) were more temperature-tolerant, as evidenced by their ability to withstand a peak compost temperature of at least 62°C (maximum 74°C) and a 21-day composting period.

Fayolle et al. (2006) demonstrated the total eradication of *P. brassicae* in bench- and large-scale composting processes at temperatures as low as 50°C, albeit, given that the moisture level was adequate (60 percent w:w). The total eradication of *Clavibacter michiganensis* was achieved by composting the plant remnants that had been contaminated (the causal agent of brown rot of potato and bacterial canker of tomato). Additionally, the finished compost proved suppressive to the pathogen. Only closed bioreactors can completely inactivate some plant viruses, most notably the tobacco mosaic virus, which is resistant to composting processes with a high-temperature tolerance (70°C). Recent studies found that it was too difficult to specify the conditions for the elimination of some additional viruses, such as the Potato spindle tuber viroid. By composting for 8 days in an environment where all materials have attained a minimum temperature of 50°C throughout that time, the nematodes of the species Globodera that cause cysts on plant parasites may be eradicated. Studies on the survival of *Escherichia coli* (particularly strain 0157), *Salmonella enterica, Listeria monocytogenes*, and *Campylobacter* spp. in composts, particularly those based on animal by-products, have been conducted in particular. *E. coli* O157:H7 and other pathogens in cow manure can be decreased using minimally managed composting procedures. When using a variety of composting systems and a large variety of source materials, compost temperatures can be reached that are greater than 60 °C. However, there is not enough knowledge on the pathogen survival hazards in cooler areas of the compost, especially when using in-vessel systems when the compost is not stirred. Safety concerns regarding the use of composted materials as growing media were raised after Steele et al findings in Australia that composted bark, which is mostly based on *P. radiata*, is a source of Legionella infections (Steele, 1990). The main pathogen of concern was *Legionella longbeachae*, with more cases of infections from this species than from *Legionella pneumophila*. They both have the potential to produce symptoms similar to pneumonia. In fact, growing media are now taken into account while estimating the risk of pneumonia in Australia. In Australia, 33 out of 45 samples of commercial potting mixes tested positive for Legionellae, according to Steele et al. (1990). Studies on the bacterium's ability to multiply in these media did not reveal increases in the amount of Legionellae in the manufactured goods, but the bacteria were able to endure temperatures between 4 and 35 °C for 6 to 10 months in the media. *L. longbeachae* has been linked to growing media in Scotland and elsewhere, linked to the use of peat diluents, possibly made worse by long storage periods allowing these amoeba-borne pathogens to multiply. Steele et al. (1990) examined two samples of peat-based growing media from Europe for Legionellae but were unable to detect these. The risk to grow media users is also regarded as low because the disease incidence in Scotland is thought to be very low and because the presence of Legionella species in growth media does not correspond with the number of cases there that have been confirmed.

Compost has been examined as a growth media component for vegetable transplant production and nursery stock (2001). If fertilization is changed, Pronk (1995) discovered that composted vegetable, fruit, and garden waste can replace up to 15% of the peat in the potting medium. Higher compost percentages led to high pH values. Prasad and Carlile (2003) reported that many species, including tomatoes, pot plants, as well as a variety of ornamentals, grew well at amendment levels up to 40% (v/v) in a thorough review of research conducted over a decade on the topic of amending peat media with composted materials based on green wastes and brewery by-products. Test species' germination, including that of tomato and antirrhinum, was largely unaffected by compost inclusion levels up to 25%. Higher rates of inclusion led to the observation of enhanced growth in many plants, which was attributed to higher concentrations of macro- and micronutrients derived from composted materials. When CGW was administered at a rate of 50%, N immobilization was identified as an issue and a decrease in the fresh weight of pot plants. The main issues with using compost in place of peat are elevated pH and salt levels, phytotoxicity from insufficient composting, and a lack of accessible N if the source materials are not appropriately prepared. A growing medium's ability to hold water is also impacted by the amount of compost it contains. Despite the foregoing, it should be emphasized that using current knowledge, compost can be produced that, in some situations, can completely replace peat moss in growing media, with comparable yield outcomes.

Anaerobic Digestates

In a controlled environment, anaerobic digestion (AD) involves the microbial breakdown of biodegradable materials without the presence of oxygen. Food scraps, animal manures, green compost, as well as maize and grass silage, are all acceptable biodegradable resources for use in AD. After proper composting, AD yields residual digestate that has been evaluated as a potential component of growth medium and biogas that can be utilized to generate renewable heat and energy (Raviv et al., 1987). Animal by-products laws, which mandate that all materials be pasteurized, govern the manufacture and use of these digestates in the majority of nations. When it comes to slurries that can include zoonotic infections, this is extremely crucial. Composting is a method that can be used to turn fibrous solid waste from anaerobically digested materials into a possible component for growing media. In their 2011 study, Ridout and Tripepi looked at the use of composted digestates as a part of growth media. High concentrations of ions were also observed, together with high pH (.7) and EC (.5.0 dS/m) values (K, P, Na, and Cl). Nitrogen contents in digestates can be very high. In the overall digestate or in the separated fiber of anaerobic digestate, the ammonium nitrogen (N) component (mineral N) accounts for up to 50% to 75% of the total Kjeldahl nitrogen. Organic nitrogen is the name given to the leftover nitrogen. If the ammonium portion is not stored appropriately, such as being covered by a membrane, it is prone to volatilization.

While ECs were lower at 1.1–1.3 dS/m, Crippa et al. (2013), using non-composted digestates from cattle and pig manures, likewise found high pH values (7.7–7.8). Dry bulk densities of 78105 kg/m3 and air contents of 43 to 50 percent with moderate water-holding capabilities were among the physical characteristics. Digestates were shown to substantially limit both the development of lettuce (*Lactuca sativa* L.), which served as a bioassay for phytotoxicity, as well as the germination of cress (*Lepidium sativum* L.), however, at a 25–30% (v/v) mixture with sphagnum peat, both outcomes were adequate. In peat that was mixed with anaerobically digested fiber at a ratio of 50%, MacConnell and Collins (2009) described the excellent growth of various kinds of nursery stock. According to Lamont and Elliott (2016), anaerobically digested dairy fiber can replace peat for poinsettia development up to 50% of the time and contains a sizable amount of nutrients. Ponchia et al. (2017) discovered that non-composted anaerobic digestates from fruit and wine distillery wastes had a pH of 6.7 and an EC of 1.41 dS/m. They also discovered that when the digestates were added at a 20 percent (v/v) ratio to mixtures of green compost and peat, Rosa and Abelia 3 grandiflora Rehd. showed improved growth. By combining anaerobic digestates with bulking materials like wheat straw and vine prunings, Bustamante et al. (2013) created composts. The composts displayed an appropriate level of physical stability, maturity, and stability for their potential use as growing media. The bulking agents had a beneficial effect, decreasing the EC and N losses during composting and also reducing the heavy metal levels in the end products. Additionally, the composts slowed the spread of the melon fusarium wilt, *F. oxysporum f. sp. melonis.*

Biochar

Biochar, or charcoal in essence, is typically made from organic waste using pyrolysis technology at temperatures between 400°C and 700°C where oxygen is either absent or depleted. Biochar is essentially charcoal used as a soil or growing media amendment. In addition to producing biochar, the pyrolysis process also yields syngas and bio-oil, two additional compounds that could be used commercially as energy sources. Biochars have been successful instruments for waste management, soil remediation, and may also enable the control of greenhouse gas (GHG) emissions through carbon sequestration. Biochars vary depending on the feedstock, temperature, and residence duration. One type of biochar might not be good for all growth circumstances and crops due to the wide variety of biochars. In South America (Colombia), materials that are similar to biochar, like burned rice husks, are often utilized to grow cut flower species. Many biochars have a total porosity between 90% and 95%, which is comparable to peat. The size of the material's particles has a significant impact on air space and EAW. Bulk density ranges from 100 to 300 g/L, which is comparable to H2-H4 peat. The physical qualities of hydraulic conductivity and water availability were shown to be improved by pelletized biochar when it was evaluated as a growth medium, however, the impact on porosity was inconsistent. According to Zaccheo et al. (2013), adding biochar to peat improves the physical characteristics of growing media, such as the air content and shrinkage. In comparison to CGW alone, Zhang et al. (2014) discovered that the addition of biochar decreased bulk density and increased porosity. Biochars frequently have an extremely high pH. Nevertheless, this might make it possible to replace all or some of the dolomitic limestone that is often used to neutralize peat. In their 2014 study, Zaccheo et al. looked at the possibility of using pinewood-derived biochar, which has a pH of 10.2, to neutralize peat, which has a pH of 4.2. The pH might be changed to a somewhat acidic value most effectively with proportions of 30% and 40%. The strongest neutralizing power was found in the fine fraction of 0.3 mm.

When biochar is used, acidic peat-based growth media that are typically limed to alter the pH for various crop demands may function without lime. The EC of biochar is often high or very high when compared to other organic substrates. Given that the former have relatively lower ECs than the latter, feedstocks including wood chips and other wood wastes may be more acceptable as candidate substrates. Additionally, because of the extraordinarily high ECs, biochar generated from agricultural wastes from glasshouses would not be acceptable as a part of growing media (Blok et al., 2017). With the exception of K, all biochars have low extractable macronutrient values; nonetheless, some have substantial Cl and SO4 concentrations (Blok et al., 2017). As peat diluents, Prasad et al. (2018) assessed four commercial biochars made from various organic sources (mostly wood) and different nutrient contents. Low quantities of N and P and often high levels of K were found in the biochars. Biochars have exceptionally high structural stability against microbial decay due to their low OUR values, which ranged from 1.1 to 4.4 mmol O2/kg OM/h. The levels of CAT extractable N (including nitrate) and P were found to be significantly reduced in biochar mixtures, whereas EC, pH, and K levels increased with the addition of biochar. These materials were mixed with peat at 10%, 25%, and 50%, and nutrients were added to bring them to the same level of nutrients as in fertilized peat. The utilization of other substances used to dilute peat, such as CGW, is significantly impacted by a biochar-mediated fall in EC. Due to the high EC, the rate of peat dilution with CGW is constrained; however, the addition of biochar opens the door to larger rates of peat dilution with green compost.

According to Blok et al. (2017), biochar made entirely from beechwood chips was somewhat phytotoxic. However, Prasad et al. (2018) discovered that utilizing cress to produce roots was frequently superior to using peat. Biochar can be used to successfully dilute peat, as shown by tomato plants growing in pots with the addition of biochar. Plant biomass was in some cases better than peat alone. Adding either biochar had no effect on the dry weights of tomato plants after 4 weeks of growth, but significantly increased plant heights in all of the biochar treatments, according to greenhouse experiments using marigold (*Tagetes erecta* L.) and tomato (*Solanum lycopersicum* L.) plants grown in 3.0-L pots. With the exception of the 5 percent wood biochar, marigold plant heights were higher than the control, with minimal impact on plant dry weights. These findings suggest that at these levels, both pelletized straw and wood biochar would be adequate to substitute peat in soilless surfaces (Vaughn et al., 2013). The cultivation of *Calathea rotundifolia* Poepp., Endl.cv. may employ biochar made from green waste as a partial alternative for peat, and tomato plants would grow better in peat-biochar mixtures than in peat alone. Its growth, particularly root volume, was enhanced by the addition of biochar and humic acid, with a mixture of 20 percent biochar and 0.7 percent humic acid mixed with CGW proving to be the most effective. Thus, when combined with humic acid, charcoal, and CGW, peat could be replaced. Plants may be protected from diseases by biochar either directly or through the induction of plant resistance by altering the microbial flora of the growing medium. The addition of biochar to specific plant growth media causes changes in the rhizosphere microbiota as well as chemical reactions in the plant. There are two novel growth-affecting mechanisms:

- Due to the biochar's chemical or physical characteristics, or both, beneficial rhizobacteria or fungi that promote plant growth are stimulated to proliferate.
- Many biochar compounds that are phytotoxic or biocidal at large concentrations are used at low amounts to encourage plant growth (hormesis).

Other Organic Substrates

After composting, the solid portion of animal dung can be used in place of peat in a container medium. The Nitrates Directive (Council of the European Communities, 1991) currently limits the application of animal manure in agriculture in addition to areas where peat is limited or expensive. After going through a methanogenic fermentation process, composts made from leached separated cow dung can be used in place of peat since they have a similar bulk density, air content, and water content. These materials allowed for seedling development that was comparable to peat mix. However, the volumetric water content within the range of 1–10 kPa tension is significantly lower than peat. Separated cattle dung compost exhibits porosity similar to that of peat on the water release curve. This is caused by a high amount of residual water that is kept at .10 kPa tensions and high air-filled porosity. The timing of irrigation is affected by this. Data on the storage of composted material made from slurry fiber combined with peat, slurry fiber, and sawdust as well as the material alone revealed that it is quite stable as demonstrated by the lack of changes in N content after storage for more than a year and by the maintenance of its suppressive capacity against a number of soil-borne pathogens. Additionally, scrap chipboard can be composted, processed, and used in place of peat in growing media. Urea formaldehyde may make up to 10% of this substance. It was discovered to have poor storage qualities as a result of N being released, which led to a low pH and large quantities of NH_4N and NO_3N.

Stability of Organic Substrates

Plants are cultivated in soilless production over periods of time ranging from a few weeks to several years. Media that is sold through retail establishments may be kept outdoors, frequently in regions that may experience considerable temperature changes. In general, it is preferred that the substrate stay undamaged while being used or even while being stored to avoid changing the growing environment in a way that will reduce productivity. However, it is significant to remember that the root zone is a microcosm where species in addition to the plant are dominant. In essence, organic growth media are living systems that may contain a great deal of biological activity that can have both beneficial and bad effects. The primary drawbacks of this biological activity include nitrogen immobilization and the substrate's physical disintegration (i.e., lack of stability); the advantages include the release of nutrient ions from slow-release fertilizers and other complex nutrient sources, as well as the prevention of disease. Crops with a long production cycle, such as woody nursery stock, can suffer from the physical disintegration of the growing medium. High-aeration crops cannot be successfully grown in moist environments (e.g., ebb and flood systems). In this instance, maintaining optimal growing conditions over the long term depends on the physical stability of the growing medium. The ability of the various peat kinds to withstand physical deterioration varies greatly. Compared to younger peats, peats with a higher degree of decomposition (H5-H6) are more structurally stable (H2-H3). The degree of breakdown was inversely related to particle size and decomposition (on the von Post scale). For coarse H5 (Irish) peat, the volume loss ranged from 5% to a whopping 47% for H2 (Finnish) peat. Over the course of the fine-grained

H2 Peat's 15-month incubation period, airspace reduced from 21% to as little as 3.5%. Volume reduction was correlated with a decrease in air-filled porosity. Peat from bogs in the Baltic States, Sweden, Ireland, and Germany of two sizes (0-3 and 5-10 mm) was compared in a different trial, which was documented by Prasad and O'Shea (1999). After nearly 2-years of incubation, the volume loss was in the range of 13 percent to 20 percent for H4 and H5 peats and 29 percent to 40 percent for H2 and H3 peats. Even worse, a volume loss of about 50% was seen in wood fibers. Changes in particle size showed that steaming of H2 or H5 peats did not appear to have a major impact on breakdown. Peat samples obtained before and after incubation revealed a trend toward lignin enrichment and a decline in cellulose in FT-IR measurement. According to Prasad (1997), coir's volume decreased by 24 percent over the course of 24 months as a result of biodegradation, demonstrating that it is a material with a fair amount of stability. Nelson et al. (2004) stated that shrinkage of 19–21% occurred just 8–10 weeks after cropping. The original lignin content of different batches of coir may have varied, and this variation in lignin content may have been influenced by the age of the coir and the fiber content. With different types of wood fiber, volume loss of between 36 and 54 percent during a two-year period was seen (Prasad and O'Shea, 1999).

In a study with coir, three wood fibers, and either H5 peat alone or in combination, Prasad and Maher (2004) discovered that the wood fibers generally degrade more quickly over time than peat. According to Prasad and Maher (2004), Toresa typically shrank far less than other wood fiber products because of variations in the lignin concentration of several materials. This investigation also discovered that coir is rather stable. The pace of breakdown of the resulting mix was reduced by the addition of H5 peat to all the non-peat elements, but the inclusion of lime sped up the breakdown. The initial lignin content had a high and unfavorable correlation with material breakdown. FT-IR spectroscopic analysis of the samples was performed before the experiment even began. According to Prasad and O'Shea (1999), there is a significant inverse relationship between substrate shrinkage and the FT-IR peak ratio 1600 cm21/1056 cm21 (lignin to cellulose). A 30% by volume addition of CGW increased bulk density and air content while decreasing pore space and readily EAW content. Monitoring height loss of the peat/CGW mixtures during a two-year period allowed for the measurement of physical degradation. H2 peat alone deteriorated considerably faster than H5 peat. When young CGW was added, both peats degraded more quickly than when adult CGW was added, which only affected the H5 peat. Additionally, in this experiment, the mixes were subjected to FT-IR spectroscopic analysis, and the peak ratio of 1600 cm21/1056 cm21, which indicates the aromatic/aliphatic ratio (lignin-to-carbohydrate ratio), revealed a negative correlation between the mixes' physical degradation. Harrelson et al. (2004) examined the physical characteristics of young pine bark and pine bark that had been aged for a year in an exposed site with and without coarse builders' sand (bark: sand 5 8:1; 11 percent sand by volume). The aged pine bark/sand substrate had a much higher container capacity and water availability than the fresh pine bark/sand substrate. The scientists came to the conclusion that due to the restricted water content, fresh pine bark-sand substrate would need frequent watering with modest amounts of water. In the course of the year, the fresh pine bark-sand substrate's accessible water content rose from 15.8 to 22.3 percent, whereas the amount of air-filled porosity fell from 36.3 to 24.9 percent. Similar results were observed in the old pine bark-sand substrate, which demonstrated a drop in air-filled porosity from 25.9 percent to 17.0 percent and an increase in accessible water from 22.7 percent to 30.0 percent. The farmers may face difficulties using both mixes for long-term production due to the decline in air-filled porosity for both mixtures over time. Biological stability is a crucial factor to take into account while managing N nutrition. There is evidence that some coir can immobilize nitrogen (Prasad, 1997). Over the course of 5 weeks, immobilization ranged from 50–60 mg N/L in some cases, to barely detectable in others. If the composting process has not been completed properly, growing media made from composted waste may contain significant concentrations of cellulose or hemicellulose, which microbes can easily break down. As a result, many peat-free media mixtures naturally have larger microbial populations and activity levels than peat-based media mixtures (Dickinson, 1995). This frequently causes a rapid loss of inorganic nutrients (microbial immobilization), especially nitrogen (Dickinson, 1995), and a change in pH at the same time. By including a buffering agent, like expanded clay or other clay minerals in the media mixes, such effects can be mitigated (Bragg, 1998). With organic media intended for retail markets, nitrogen status changes are frequent. In some cases, soluble N release from complicated inorganic and organic N sources may occur; in other cases, notably with wood and bark-based media, N immobilization may take place.

CHAPTER 3

ROOT GROWTH IN SOILLESS CULTURE

CONTENTS

ROOT ANATOMY OVERVIEW

The arrangement of the root tissue is ideal for the effective delivery of nutrients and water to the shoot. The majority of root types share a similar basic structure, while secondary-growing roots will have more evident vascular tissue. Cell division in the root apical meristem, which generates a variety of distinct root cells, causes primary root growth. At the distal side, the apical meristem is shielded by numerous layers of root cap cells. Additionally, mucilage is produced by the root cap, which aids in root elongation. At the root tip, root cap cells slough off and die, and new cells made by the apical meristem take their place. In the elongation zone just proximal to the apical meristem, other root cells produced by the apical meristem lengthen and widen. Following this, root cells begin to differentiate in the differentiation zone, which is close to the elongation zone. A layer of epidermal cells, some of which have root hairs, sometimes an exodermis, multiple layers of cortex cells, a layer of endodermis cells where Casparian strips start to develop, the pericycle, and vascular ground tissue with phloem and xylem are formed as cells differentiate from the outside in (See the figure below). The maturation zone is where these tissues develop into fully functional tissues. The degree to which each of these cell types is present and/or modified depends mostly on their immediate environment and, to a lesser extent, on the carbon flow to the root. Each of these cell types has distinct tasks. The ratios of auxin (which promotes cell division), cytokinin (which promotes cell differentiation), and gibberellin (which promotes cell maturation) as well as the species and environmental conditions have a significant impact on how long each of these zones (cell division, elongation, differentiation, and maturation) lasts (which inhibits cytokinin signals). When bacteria and fungi in the rhizosphere come into touch with epidermal cells, they exude acids and carbon-rich chemicals and create root hairs, which increase the absorptive root surface area and the amount of substrate or soil that the root system explores. In some species, the inner root surface may be protected from dehydration by a well-developed exodermal area (Hose et al., 2001; Lynch et al., 2014), while specific passage cells permit the water to continue flowing.

The primary structure of the root, the cortical layer, is where metabolic processes and interactions with symbiotic bacteria or fungi take place. Aerenchyma or other types of root lacuna may develop in this layer as a result of controlled cell death, protecting the roots from drying out, enhancing oxygen diffusion, and lowering total tissue respiratory costs. The storage and metabolism of carbohydrates both heavily involve the cortex. The endodermis is the following layer, and this is where the root may regulate the diffusion and uptake of nutrients and water into the stelar tissue. The pericycle, a single layer of cells with the founder cells that start lateral root production, is located right inside the endodermis. The vascular tissue that transports water, nutrients, and sugars between roots and shoots is found in the stele, which is the core of the root. The xylem is loaded into roots, which are often very active organs. Phloem activity in roots is typically dominated by chemicals translocated from the shoots. Many dicotyledonous plants' roots experience secondary growth after primary growth, which increases root diameter and results in the production of additional vascular tissue. The role of the root within the framework of the overall root system has a significant impact on the rate of secondary xylem development. Environmental and genetic factors influence the investment in tissues that transport water. Even under the same, ideal growing conditions, the roots of six distinct genotypes of soybean displayed significant variance in root hydraulic conductivity to water, with changes in the surface area of the stele accounting for 35% of that variation. Significant variations were also seen in axial conductance and cortical volume among these genotypes. Primer cells in the pericycle, where the initials that will produce a new root meristem are triggered, are the source of lateral root growth. To get to the rhizosphere, the freshly formed root must then push through pre-existing endodermal, cortical, and epidermal tissue. The new root can push through existing cells thanks to auxin-induced acidification of cell walls, which weakens the connections between cell walls.

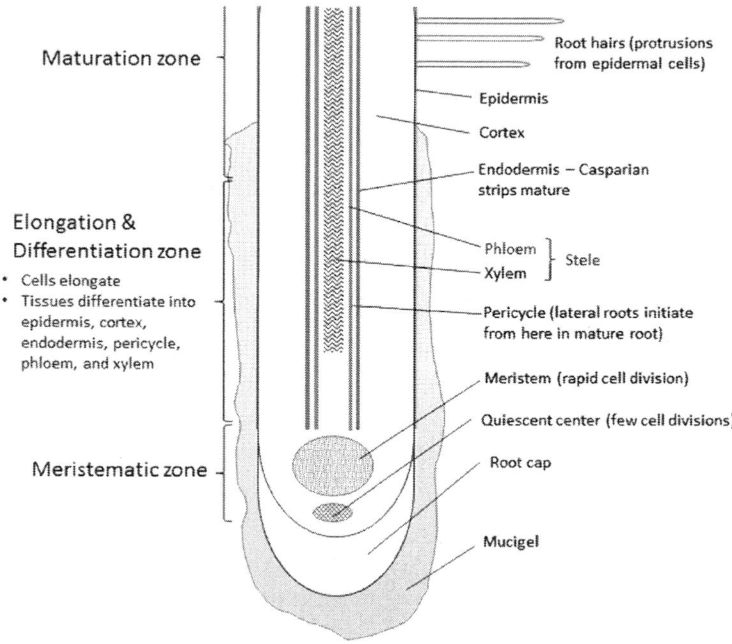

Figure 12: Root parts (longitudinal section).

Source: Taiz and Zeiger, (2010).

ROOT PHYSIOLOGY IN SOILLESS CULTURE

When roots are growing in soil, they are subject to shifting soil particles, decomposing or herbivorous soil organisms, and variations in soil-water environment, temperature, and nutrient availability. In contrast to considerably more stable hydroponic or aeroponic environments, roots tend to grow more defensive tissues in reaction to their external environment under these circumstances. Increased lignin and suberin content in epidermal cells, which reduces the effect of soil strength on the roots, are modifications in these protective tissues. Under these circumstances, roots also tend to have a somewhat lower length-to-mass ratio (i.e., are denser). Shallow roots brown more quickly than deeper growing roots, suggesting that disturbance and changing environmental factors contribute to the pace of browning. When roots turn brown, their metabolic activity drastically decreases, likely as a result of significantly decreased active nutrient uptake, which can account for as much as 50% to 70% of the root respiratory costs for unsuberized roots in the case of both fast- and slow-growing herbaceous species. When roots are submerged in aerated nutritional solution continuously, they develop more length per unit of root mass while spending very little on protective tissue or supporting tissue, maximizing root-water nutrient interactions. The nutritional solution maintains the root mass, thus the latter is not required. However, when the root mass is periodically disrupted by bending or flexing, the roots exhibit a noticeably increased tissue density and rigidity. When used correctly, aeroponic techniques provide the roots with plenty of oxygen, moisture to prevent dehydration, and nutrients to meet plant needs. When roots are suspended in the air, more supporting tissue is required to maintain structural integrity; hence, compared to hydroponics, significantly more biomass is invested in cortical and epidermal cells. The typical browning of roots caused by epidermal cells accumulating suberin and lignin under soilless conditions of modest disturbance is absent or delayed, and it is likely that these roots are still considerably more metabolically active.

ROOTS AND SOILLESS ENVIRONMENT

The availability of carbohydrates to the root has a significant influence on root development. Since it was first proposed in the 1960s, the concept of a functional equilibrium, in which newly assimilated carbon is preferentially allocated to organs that acquire the most limiting factor to plant growth, has had a significant impact on how we generally understand carbon allocation patterns and tradeoffs. Plants spend a greater portion of their carbohydrates in the root system when nutrients or water are scarce, presumably because these tissues become stronger carbon sinks when shoot growth is constrained by nutrient and water availability. Because the nutritional requirements of root tissue are lower than those of shoots, roots can continue to grow even when nutrient uptake is limited, whereas shoot growth stops, increasing the root mass fraction. By directing carbohydrates into root development, plants will be able to absorb more nutrients, perhaps reducing the effects of a limited nutrient supply. High CO_2 and high light levels encourage carbon uptake, and the availability of nutrients may limit biomass production. When nutrients become scarce, a typical response is an increased allocation of carbon to the root system, or an increase in root mass fraction in comparison to plants grown under ambient CO_2 or low light circumstances. The allocation patterns drastically shift to investing more mass in shoot growth in soilless settings, when plants are frequently given a constant supply of water and high nutritional conditions, resulting in root mass fractions as low as 0.05. Environmental factors have an impact on root anatomy in addition to the harmony between root and shoot growth. For instance, when eight different grass species were shaded from 30 percent of the sunlight, the proportion of root stele tissue increased and the size of the xylem vessels increased, indicating a greater investment in water transport capacity to support a proportionately greater investment in transpiring leaf area.

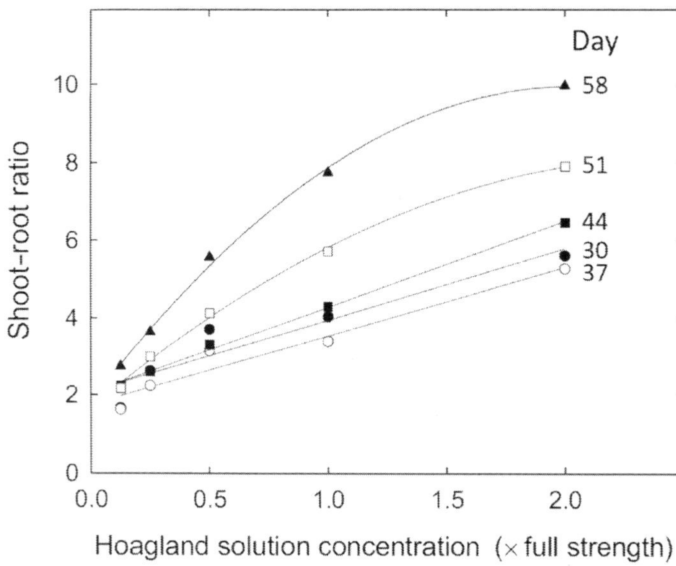

Figure 13: Impact of nutrient solution on the root-shoot ratio (Scarlet sage).

Source: van Iersel (1997).

The effects of temperature on root development are poorly described. In general, when the temperature rises, root respiration increases. Increased metabolism, however, does not always translate into faster rates of nutrient intake or even faster root growth. According to the native habitat of the species, the effects of root zone temperatures on root growth have been evaluated in a number of articles. For instance, it was discovered that alpine plants can survive at temperatures as low as 0.8°C to 1.4°C, whereas polar semidesert plants may grow to some extent at 1°C to 3°C but grow best between 12°C to 20°C in the soil. Temperatures between 20°C and 25°C seem to be ideal for root growth in more temperate species like walnut trees. When allowed to adapt to soil temperatures, Concord grape roots showed their peak rate of respiration (a proxy for metabolic activity) around 25°C, after which time respiration decreased and membrane leakage sharply increased. According to this study, there is a direct correlation between soil water content and membrane leakage. While roots grown in soils with less than 5% water content showed dramatically increased membrane leakage and decreased respiration, roots grown in adequately watered soils at 25°C showed very little membrane leakage. When soil temperatures rose from 35°C to 50°C, *Dichanthelium lanuginosum* showed significant decreases in stomatal conductance and carbon gain. Negative impacts could only be partially countered by increasing soil water content, indicating that decreased water uptake was causing plant water stress. Temperature had an impact on root respiration for more drought-resistant species like citrus, but short-term low water availability had little effect.

It is challenging to determine the precise ideal temperatures for root growth since seasonal phenology and water availability can contradict the impacts of soil temperatures with field research documenting root growth. The distribution of roots of field-grown plants may be affected by temperature gradients in the soil, with winter root growth continuing better in deeper soil layers with mild temperatures than in colder surface layers. A consistent root zone temperature of 20°C produced more leaves, more leaf area, and more dry weight in pepper plants (*Capsicum annuum*) cultivated aeroponically in greenhouse research when compared to plants having their root zone at ambient temperatures (which fluctuated between 25°C and 40°C). Root zone temperature can obviously have a significant impact on shoot growth, even if the effects of the various root zone temperatures on root growth were not assessed. Such effects might be far more pronounced in soilless culture than in crop cultivation on soil. Temperature fluctuations in the root zone can be reduced in soil-based production due to the soil's substantial thermal mass. In soilless production, this temperature buffering effect is significantly less noticeable. In fact, root zone temperatures in container-grown plants can vary much and quickly, particularly if the pots are in direct sunshine. When plants are cultivated in black plastic pots, it is easy for root growth responses to high substrate temperatures in containers to inhibit root growth at root zone temperatures of about 38°–40°C for 6 hours each day. Root cell membranes are irreversibly damaged after a brief (20–30 minute) exposure to root zone temperatures between 46 and 52 degrees Celsius. The majority of the energy that heats up container substrates originates from the side walls of the containers being exposed to radiation; larger containers have greater center-of-pot buffering capacity than smaller containers. Media in pots with a larger surface area showed higher media temperatures than media in pots with the same volume but a smaller surface area, indicating that the geometry of the container also affects the ability to buffer temperatures. The root zone temperature is momentarily lowered by irrigation with water that is cooler than the root zone, but the water must be administered in such a way as to completely saturate the container in order to "flush" away the heat energy. Lower water content caused the surface to cool due to evaporation. However, when less than saturating amounts were used, the temperature at the container's core rose (Martin and Ingram, 1991). When the irrigation water's temperature was higher than the media's, the result was the opposite (Heller et al., 2015).

Figure 14: Root heat stress damage on potted plants.

Source: Ingram at al., (2015).

Roots are affected differently by high root zone temperatures even when they are exposed to non-damaging substrate temperatures. With rising temperatures, root respiration rises exponentially, resulting in greater medium CO2 concentrations and increased oxygen demand. Temperature and temperature gradients have a significant impact on root growth, and the ability of roots to adapt to higher substrate temperatures is greatly influenced by species. For instance, diurnal root extension rates in hydroponically grown maize were 50% greater at 26°C than at 21°C and constant when the temperature was held constant (Walter et al., 2002). This study also revealed two expansion zones that demonstrate varying temperature sensitivity inside the elongation zone. Even in stable conditions, their 3-D approach of root observation allowed them to see regular (40–70 minute interval) oscillations in root-tip development direction, which may help to explain why many roots have a corkscrew-like look. Within an hour of changing the temperature from 21°C to 26°C, new steady state growth rates were observed, demonstrating that root growth may quickly respond to and adapt to changes in the thermal environment. The nitrogen supply, either ammonium or nitrate, has a significant impact on the relationship between the temperature of the growth media and root growth and development. A controlled experiment with barley showed that roots exposed to a gradient from 20°C to 10°C had strongly accelerated development and after 30 days these plants had approximately 3.5 times greater shoot and root biomass than plants exposed to a uniform 20°C root zone temperature.

The temperature gradient itself can also have a significant impact on plant and root growth. Aeroponic and hydroponic systems often do not have these temperature gradients, however, potted plants may experience substantial horizontal temperature gradients in addition to vertical gradients. For roots to grow, water must be present in the root zone. A root exposed to water stress will often have smaller cells and denser tissue because newly divided cells cannot absorb enough water to exert the pressure necessary for cell elongation. The fact that soil strength increases when soils dry out and raises the pressure required for roots to break through the soil is a confusing issue for plants that are growing in soil. It is simpler for roots to push soil particles aside and push through moister soil because the friction between soil particles and between soil particles and the surface of the roots is reduced. It is highly unlikely that substrate strength will considerably hinder root extension due to the considerable porosity of soilless substrates. As most ions migrate to the root surface by bulk flow or diffusion, maintaining adequate soil-water availability also increases nutrient uptake by the roots. Transpiration by the shoots determines the rate of bulk soil water flow to the roots. The mass flow of water and dissolved nutrients to the roots will be reduced if low soil-water availability restricts transpiration.

The concentration gradient from the bulk soil solution to the roots and the tortuousness of the diffusion channel determine how nutrients diffuse. Lower soil-water content results in increased tortuosity, decreased diffusion, and reduced nutrient transport into the rhizosphere. It's interesting to note that with soilless substrates, root growth itself reduces tortuosity, probably by bridging the pore holes inside the substrate. In contrast to gravitropism, roots respond to moisture gradients by growing across short distances in the direction of moister soil parcels. According to Bao et al. (2014), hydropatterning in response to water gradients determines the spacing between founder cells in the pericycle, which in turn determines lateral root density. Root branching patterns and growth direction are also strongly influenced by water availability and its gradient. Water shortage does enhance the relative carbon allocation to the roots, reducing the shoot-root ratio even though it generally results in smaller plants and overall smaller root systems. Research findings on Arabidopsis using labeled carbon have demonstrated that transporters involved in phloem loading and unloading, as well as sucrose transport in the phloem, are elevated in response to drought (Durand et al., 2016). According to this study, Arabidopsis roots respond to a water shortage by storing much more sucrose, glucose, and fructose, which increases osmotic pressure in the roots. Under drought conditions, this ought to help with water uptake and cell expansion. On the other hand, roots can quickly run out of the oxygen required to continue cellular respiration when all pore spaces are completely filled with water. When oxygen is present, roots can perform oxidative phosphorylation, producing 60 adenosine triphosphate (ATP) molecules for every sucrose molecule. Low oxygen levels render oxidative phosphorylation ineffective, and the only pathway that produces the maximum amount of ATP—4 ATP per sucrose—does so through glycolysis—a 93% drop in respiratory efficiency. Such conditions provide an immediate negative impact on the two energy-intensive processes in a developing root system: cell division, which requires ATP to sustain H 1 gradients across cell membranes, is significantly curtailed, as is active food uptake. Therefore, it should not come as a surprise that many overwatered plants exhibit significant nutritional deficiencies and that the issue with such plants is frequently misdiagnosed. Ethylene can also be produced in low oxygen circumstances. In response to increased transport of 1-aminocyclopropane-1-carboxylic acid, an ethylene precursor, to the shoot where it was converted to ethylene, tomato plants overwatered to low oxygen levels displayed leaf symptoms very similar to withering.

Contrary to popular belief, a lot of plant species close their stomata in response to low oxygen levels even while there is ample water available for transpiration. Surprisingly, increasing O2 conditions in the root zone had little impact on plant growth; tomato plants only displayed lower growth when levels reached 30 mg/L or higher (relative to 8.5 mg/L ambient). Such circumstances are unlikely to exist on Earth, but they may be crucial when considering the controlled growth of plants in space, where they can enhance oxygen levels and lower CO2 levels. Because root respiration depends on oxygen, hydroponic solutions must be appropriately aerated to avoid hypoxia. The high porosity of the substrate makes it easier for oxygen to diffuse into and throughout the substrate when plants are grown on soilless surfaces. As a result, the amount of water in the substrate has no impact on the partial pressure of oxygen in the substrate pores, and even extremely wet substrates showed no signs of hypoxia (van Iersel and Dove, 2013). Their findings do not, however, exclude the possibility of hypoxic conditions in the rhizosphere of roots, where oxygen demand is highest. Plants typically allocate significantly more biomass to the shoot when exposed to constant high nutrient conditions (such as hydroponics) (van Iersel, 1997). When drip irrigation is utilized to deliver nutrient solution to a particular area of the substrate or when controlled release fertilizers are used, localized nutrient distribution can happen in soilless substrates. The patterns of root allocation for the entire plant can be overridden by localized nutrient availability. If resources are very mobile and lateral roots compete with one another for nutrition, having a dense local lateral root system may be detrimental. Plant life may depend on the root system's flexibility, or its capacity to react quickly to shifting water and/or nutrient availability. More recent research has demonstrated that low phosphorus circumstances increase lateral root length while decreasing the root length to the first lateral root and the space between lateral roots. The reactions to nitrate are comparable. Localized high nitrate circumstances have been shown by Zhang and Forde (2000) to promote lateral root branching. Localized high auxin levels in the roots modify this response. High internal nitrate concentrations, however, suppress the signal that promotes lateral root development. Arabidopsis exhibits very responsive root hair development to nutrition availability. Root hair density and root hair lengthening both occur under low P circumstances.

Under low P circumstances, the ratio of sucrose to glucose decreased while the pools of both glucose and sucrose grew. Cell elongation and root foraging may be facilitated by the subsequent rise in osmotic pressure. In comparison to crop production in soilless substrates or hydroponics, where fertilizer application rates and nutrient availability are often high, such root flexibility is probably more significant in natural ecosystems and low-input agriculture. The size, characteristics, and subsequent response to shoot growth of the root system are greatly influenced by the physical space available in the root environment to allow root system expansion. The ranking of root characteristics and CO2 responsiveness was influenced by the size of the container, according to a recent study specifically designed to determine whether wheat responses to elevated CO2 conditions were consistent among genotypes cultivated under greenhouse conditions in two different pot sizes. In general, growing plants in containers can hinder proper root development due to space constraints, unnaturally high temperatures in the root zone, altered temperature gradients, accentuated soil-water extremes, and altered textural and structural properties of the growing medium in comparison to natural soils.

Figure 15: Root development under different pot sizes.

Source: homesteadandchill.com/potting-up-seedlings/

Root girdling and other structural root deformities can result from root space restriction in annual and perennial plant species, but it can also significantly slow down shoot growth (Hsu et al., 1996; van Iersel, 1997), which might be viewed favorably if containerized production aims to dwarf trees (Hsu et al., 1996). Root limitation improved root and leaf N concentrations and leaf photosynthetic rates while decreasing net nitrate uptake in grapevines. This implies that nitrate efflux and/or generally decreased root surface area in these vines were not offset by upregulating absorption rates at the root level, as was also observed in tomato. After being sold, many plants that were grown in pots are then moved into bigger containers, gardens, or farming fields. Although research trial outcomes have been mixed, root limitation during transplant formation can affect posttransplant performance. The degree of root restriction-induced posttransplant growth decreases is likely influenced by the species, container size, and growing environment. Circling roots may form inside a container when woody plants are grown in an overly small container or are potted up in a way that twists or compresses the root system (see figure below). It is possible to kink, J-root, twist, etc. these roots. When perennials are transplanted in the field, these roots, which will eventually form the permanent structural roots of the root system, may begin to girdle the tree trunk, resulting in unstable trees and faults that might not be noticeable for some time after the tree was transplanted. Root circling in potted plants may be minimized with the use of root pruning pots, which use an air layer to limit root growth, or copper-treated pots, but the effectiveness of copper-treated containers depends on the capacity of the container.

Figure 16: Girdling root on a tree.

Source: www.reviewjournal.com/

Other pot designs that aim to lessen structural roots' circling and promote fine-root growth include air pots, decomposable pots, and fabric-based pots. In the first year following their transplantation into the field, red oak trees grown in root-pruning, air-potted, or fabric containers had reduced water stress than bare-rooted plants, indicating an advantage of these fine-root stimulating pots. These advantages, however, were not apparent in drier urban environments or in the years that followed. Root properties within a species are far more influenced by the environment than by genetics. In a recent study employing 24 distinct genotypes of wheat in 12 different soil settings, it was discovered that genotype only described 12% of the variation in root depth, environment explained 48% of the variation, and genotype by environment interactions explained 40% of the variation (Acuna and Wade, 2012). Given the soil penetration resistance varied significantly between sites, it makes likely that the environment has a significant impact on root depth. However, features that caused genotypes to have deeper rooting depth at one site did not always foretell rooting depth at a different site.

ROOT DEVELOPMENT IN SOILLESS CULTURE

Fine roots [first and second-order lateral roots, counting from the outside to the interior] make up the majority of the length of the entire root system. In order to increase the amount of root surface area in contact with the soil solution and the total uptake of water and nutrients like N and P, fine roots can enlarge into smaller soil pores. The primary root's growth is largely indefinite, in contrast to the determinate growth of lateral roots and tertiary roots. Primary roots have a significantly higher gravitropic response than secondary or tertiary roots, whose responses can also be influenced by drought stress. Drought stress causes roots to develop at steeper angles. The effectiveness of N acquisition is also significantly influenced by root growth angles and branching patterns. In diverse ecosystems, where branching patterns are closely regulated, branching occurs more frequently when roots come across advantageous patches. The diameter of the roots closest to the surface tends to shrink with aging. Although in many species the first 1-3 orders may not differ significantly in diameter, roots of the lowest order tend to have the smallest diameter. Natural soil requires root systems to continually explore new soil areas and regenerate as depletion zones emerge. There are few studies on fine root turnover in soilless systems, and neither root longevity nor turnover rates in hydroponic or aquaponic systems have been estimated. The endodermis and exodermis are formed from the cortex's inner and outside halves, respectively. The exodermis and endodermis are key regulatory structures for water and nutrient intake from a physiological perspective. In the exo- and endodermis, older root system components and aged determinate roots generate suberin layers where water flow with dissolved nutrients is compelled to join a symplastic pathway.

Exodermal and endodermal tissues go through three stages of maturation. Stage I involves the development of Casparian bands between endodermis cell walls, stage II involves the deposition of suberin lamella between individual endodermis cell membranes and cell walls, and stage III involves the strengthening of the cell walls through the incorporation of cellulose, lignin, and occasionally suberin. Suberin prevents apoplastic water flow, therefore in order for the solution to pass via the endo- or exodermis, it must cross the plasmalemma. As a result, the root system has points of control where nutrients can be actively or passively taken up into cells. Exo- and endodermis function as important barriers against pathogen entrance. Environmental factors are probably what regulate the exodermis development and the presence of lignin in the epidermis. In addition, roots thicken and tissue density rises as soil resistance increases; this is probably due to decreased cell elongation and the presence of more cell wall material per unit of root volume. Roots in soilless systems typically face relatively little resistance, which may lower exodermal tissue formation and lignin levels in the epidermis. The maturation zone of lateral roots is where root hair production often takes place. Although growth media has also been proven to have a significant impact on root hair density, root hairs often increase in density under low P circumstances. For instance, when peanuts were grown in loamy sand, a 4:1 mixture of coarse sand and fritted clay, or fine sandy loam in a greenhouse, 47%, 17%, or 14% of the roots formed root hairs, respectively. In Arabidopsis, where root hair production is regulated by auxin gradients, substantial research has been done on root hair creation. The amount of nutrients analyzed, external and internal concentrations, species-specific root characteristics, soil-water content, root age, and interactions between a wide variety of external and internal parameters all affect the rate at which nutrients are absorbed per unit of the root surface. Although this plasticity reduces as roots get older, high- and low-affinity nitrate transporters can swiftly up- or down-regulate in response to shifting nitrate concentrations in the microenvironment. Nitrate presence causes changes in root hydraulic characteristics, which are species-specific and probably related to internal N requirement. For instance, there was a strong positive linear relationship between nitrate uptake rate and the effect of nitrate on relative water flow in an experiment relating the effect of nitrate on water flow and nitrate uptake rate under high N conditions across seven species with varying relative growth rates.

ROOT NUTRIENT UPTAKE

The fundamentals of the nutrient solution and its administration remain the same even though different methods are employed in soilless culture and unique nutrient solutions that are appropriate for each system are frequently presented. Extremely, the cost and usability should be taken into consideration when selecting the cultivation system. To get the best yield and quality, a system must be selected and a fertilizer solution that is appropriate for the plant type or growth stage used. In other words, even in a controlled environment, good plant cultivation depends on an understanding of how to regulate the nutrient solution. There are 92 known natural elements, and 60 of them are present in plants. Many of these substances are not thought to be necessary for plants, though. The below listed qualities define an essential component:

- The plant exhibits symptoms and is unable to complete its life cycle if the element is missing.
- Deficit symptoms vary depending on the absence of particular components,
- Other elements cannot replace the element in the function; only the element may cure the ailment.
- The element is directly connected to plant metabolism or nutrition.

17 elements have been identified as needed for higher plants, according to recent studies on plant nutrition. These elements are separated into two major categories in accordance with the quantities needed by plants. The macronutrients (main elements) are those that must be consumed in relatively large amounts, and the micronutrients are those that must be consumed in relatively tiny amounts (minor elements, trace elements). Nine components make up the macronutrients:

- carbon (C),
- oxygen (O),
- hydrogen (H),
- nitrogen (N),
- phosphorus(P),
- potassium (K),
- calcium (Ca),
- magnesium (Mg), and
- sulfur (S).

The micronutrients include eight elements:

- boron (B),
- iron (Fe),
- manganese (Mn),
- zinc (Zn),
- copper (Cu),
- molybdenum (Mo),
- chlorine (Cl), and
- nickel (Ni).

Since water makes up the majority of fresh vegetables, just around 10% of the initial fresh weight is made up of dry matter. About 6 percent hydrogen, 40–45 percent oxygen, and 40–45 percent carbon makes up the dry matter (H). It is typically not necessary to think of the provision of carbon, oxygen, and hydrogen as fertilizers because these important elements are delivered to the plants as carbon dioxide and water. Therefore, only six elements (N, P, K, Ca, Mg, and S) are categorized as macronutrients after removing these three elements. K>N>Ca>P>Mg is generally the sequence in which macronutrients are absorbed by plants, and N and K concentrations in vegetable leaves are often larger than those of other minerals, though the ratio varies depending on the vegetable species. When deciding on the ideal nutrient solution composition and how to control the water and nutrient supply, the mineral levels are crucial. Vegetables also require more calcium than other plants do. In either scenario, when there is insufficient natural supply, the vital components must be artificially given to make sure they are enough for the plant's demand.

The term "beneficial elements" originally meant substances that are necessary for a particular plant species or that, when present, have a growth-promoting impact. Aluminum (Al), sodium (Na), silica (Si), cobalt (Co), selenium (Se), and silica are thought to be advantageous components for higher plants. However, "beneficial ingredient" can also signify something else. The Japanese Ministry of Health, Labor, and Welfare has established suggested dietary levels for 13 different mineral types, including selenium (Se), chromium (Cr), and iodine (I). Additionally, bromine (Br), fluorine (F), and rubidium (Rb) are all classified as necessary micronutrients for humans and are found in quite high concentrations in the human body (Yamada, 2010). In other words, if plants have adequate levels of these elements, humans will benefit from them. Products made in plant factories may have added value since it is simpler to control the amount or timing of supply of these ingredients in a controlled environment. The ability of plant roots to absorb water and nutrients is one of their most crucial physiological functions. Near the root tip and the younger portion, where the root is growing, nutrients are often absorbed quickly. The following two ideas are typically used to describe how plants absorb nutrients.

- **Apoplastic transport:** The first is the apoplast's route, which is made up of the spaces between cells and their exterior cell walls. Ions are carried by the apoplast (apoplastic transport) to the Casparian strip, where they are picked out and absorbed into endothelial cells through a respiration-dependent mechanism.
- **Symplastic transport:** Another way is via the ion channel, a cellular protein that regulates the entry and exit of particular ions into and out of cells. The difference between electrical potential and chemical potential is the force that propels the passage of ions over the channel; as a result, there is no energy required for the transport. The protoplasmic connection is used to transmit the nutritional ions that are taken in by the cell (symplastic transport).

These processes enable plants to choose and positively absorb only particular ions. To reach the entire plant body, the nutrients go upward with the water through the xylem. The cell wall is eventually breached by the nutrients dissolved in water, allowing the cells to absorb the nutrients. Water movement inevitably has a significant impact on the distribution of the minerals Ca and B. Water is propelled by root pressure flow at night and transpiration flow during the day, respectively. Basically, the majority of nutritional ions only flow through xylem tissue, while some of them are redistributed throughout the plant's tissues via the phloem.

All nutrients are provided to the plants in soilless culture by the nutrient solution. The nutrient solution is, in essence, a carefully produced liquid fertilizer. The nutrition solution should meet the requirements listed below:

- all essential elements for the plants are contained,
- all elements exist in ion form,
- the concentrations of each ion and total ions are within appropriate ranges for normal plant growth,
- harmful substances and pathogenic microorganisms are not

contained,

- the solution pH is around 5.5–6.5 and stable,
- the solution can be made using relatively inexpensive fertilizers,
- the concentration of each ion, ratio of ions and pH do not fluctuate greatly during the cultivation period, and
- in the deep flow technique (DFT) and nutrient film technique (NFT), enough oxygen is dissolved for root respiration.

The amount of each nutrient ion in the solution is determined by the composition of the nutrient solution. Numerous studies have been done on this subject, and numerous compositions (formulas) have been developed. The type of plant, growth stage, growing season, open or closed system, use of a substrate, type of substrate, desired product quality, and other factors should all be taken into consideration while selecting a formula. The concentration and the ratio of each ion, however, are the sole variables. A small adjustment in fertilizer management can have a significant impact on plant growth and product quality in soilless culture. As a result, effective nutrient management can be achieved by taking into account not only the ion composition and concentration but also other factors including temperature, dissolved oxygen concentration, and microbial flora.

Making the Solution

The concentration of H+ ions in the nutritional solution is indicated by the pH (potential hydrogen) of the solution. If the pH is 7, the solution is neutral. If it is lower than 7, it is acidic, and if it is higher, it is alkaline. The pH of the solution alters the ion shape and/or solubility of the nutrients, which impacts their efficacy. For instance, when the pH of the solution is between 5 and 6, more than 80% of the P is present as H2PO4-, which is the form in which plants primarily absorb P. When the pH reaches 7, P in H2PO4- form, however, drops to 30%. Additionally, at high pH, the stability of chelate Fe, such as Fe-EDTA, reduces whereas Ca readily interacts with P to produce a precipitate. Although the pH is normally kept between 5.5 and 6.5, leaf onions and certain other species benefit from a solution pH of 4.5 to 5.5 to maximize Ca and P utilization. The uneven uptake of cation and anion by plants is the main cause of pH fluctuations in solutions. When a plant takes up cations, it releases H+ from the root, and when it takes in anions, it releases OH- or HCO3. As a result, when anion uptake exceeds cation uptake, the solution pH tends to rise, and vice versa. Additionally, in substrate culture, the solution pH is impacted by the pH of the substrate. In general, inorganic substrates like rockwool and rice husk charcoal tend to raise the pH of the solution whereas organic substrates like peat and coir tend to lower it. Although NO3- is the primary form of nitrogen in nutritional solutions, nitrate reductase converts NO3- in plants into NH4+, which is then digested into amino acids. Since enzyme activity depends on the light intensity, if NO3- is the primary source of nitrogen, the supply of NH4+ tends to be lower than that required by the plant under low light intensity. As a result, NO3- and a little amount of NH4+ are typically used to encourage plant development under low light conditions. In contrast, if the rate of NH4+ absorption exceeds that of NH4+ assimilation, NH4+ builds up in plant cells and becomes detrimental as a result of aberrant N metabolism in the plant. Due to the antagonistic effects of NH4+, Ca deficiency in the plant is readily apparent. Additionally, factors such as plant species, solution pH, solution temperature, and others influence the properties for which the nitrogen form is preferentially absorbed by the plant. As a result, NH4+ will be useful in soilless culture as a growth stimulator for more plant species when:

- the solution pH is kept constant,
- the NH4+ concentration in the solution is kept lower,
- the NH4+ is supplied in proper quantities in accordance with the rate of nitrogen assimilation by the plant.

In this scenario, the actual solution volume in the cultivation system affects the amount of nutrients actually delivered to the plants. Additionally, when the concentration is controlled, some ions, such as NO3-, H2PO4-, and K+, have a tendency to be rapidly absorbed by plants, leading to the absorption of a significant amount of these ions by plants. The yield and quality of the product, however, stop improving with an increase in this type of ion's absorption after a certain point (luxury absorption). In addition, fruit vegetables grow vegetatively rather than reproductively, and this imbalanced growth has a negative impact on the plant's ability to produce fruit and set fruit. In order to prevent plants from absorbing unnecessary nutrients, a novel idea for fertilizer administration is called quantitative control of nutrients. When fertilizer application is quantified, the weight of the fertilizer is established by early measurements of the quantity of nutrient uptake by the plant and put in the nutrient solution tank once every week or two weeks. A weekly addition of N fertilizer comprising 7 g of NO3- is made to the solution tank, for instance, if the plant absorbs 1 g of NO3- each day. Ions can be vigorously absorbed from the solution under a wide variety of solution concentrations, even if the nutrient content varies as a result of this management. Additional benefits of managing nutrients quantitatively include avoiding overfertilization, lowering fertilizer costs, and decreasing environmental pollution by draining nutrient solutions into the environment because all applied fertilizer is utilized by the plants. The use of this technology is anticipated to grow as more information about plant nutrient uptake and automatic fertilizer use management systems are developed.

The structural, morphological, and physiological properties of roots are highly changeable organs that are influenced by both genetics and the environment. Despite having developed to promote plant performance in natural soils, roots may adapt to grow and work in soilless substrates such as hydroponics, aeroponics, rockwool, and substrates made of organic materials due to their plasticity (e.g., peat, coconut coir, bark). High porosity, low physical strength, and often readily accessible water and nutrients are characteristics of soilless surfaces. These elements all influence the development and growth of roots. Despite the significance of the root system, research on soilless substrate roots has lagged behind that on crops grown in soil. Improved knowledge of the root systems in soilless substrates may lead to more effective water and nutrient use.

CHAPTER 4

CHARACTERISTICS OF SOILLESS MEDIA

CONTENTS

OVERVIEW

You can use an organic or inorganic soilless medium. These mediums can be utilized singly or in combination. There isn't a single substrate or combination that works for all cultivating conditions and species. The environment, the species to be grown, the culture phase (germination, rooting of cuttings, plant production, plant breeding), and the system of cultivation should all be taken into consideration when choosing each specific material. The physical and chemical qualities of the media vary. With time and use, the physical, chemical, and biological substrate qualities can deteriorate and change, which may have an impact on crop management and behavior. The pore structure of substrates that have undergone mechanical degradation may change, which may have an impact on the retention and transport of nutritional solution as well as root aeration. This chapter's major goal is to look into the physical and chemical characteristics that are necessary for plant growth.

PHYSICAL PROPERTIES

Solid, aqueous, and gaseous phases make up every type of media. The physical traits of these three phases will be covered both separately and together in the sections that follow.

Bulk Density

The bulk density (BD) of a medium, which is measured in g/cm3, is its dry mass per unit of volume (in a moist state). The literature contains many techniques for measuring BD (as well as other physical characteristics). Some techniques are employed mostly for study (e.g., the standard ISHS method). Others are employed as industrial norms in some nations or areas of the world (e.g., BS EN 12580:2001 in the United Kingdom, both the LUFA and DIN methods in Germany and the CEN method in the EU). However, they are all founded on the same idea: moist material is allowed to settle inside or is compressed using known pressure into a cylinder of known volume. After that, it is fully dried down and weighed. Since many media consist of multiple components, the features of each component affect the medium's overall BD. These are the arrangements of the individual and combined particles, the BD of the constituents, and the compaction characteristics. Particularly, media elements with vastly different particle sizes as a blend have greater BDs. Similar to media made of similar particle sizes, they have less total porosity (TP), water-holding capacity, and air-filled porosity (AFP). The choice of media is impacted by BD in a number of ways. For instance, a high BD medium is needed for outdoor tree seedling cultivation to prevent container instability in windy conditions. This can be accomplished by adding heavy mineral components to the mixture, such as sand, soil, clay, or tuff. On the other hand, high-intensity greenhouse crops need media with low BD because they often get irrigation and risk oxygen deficit if hydraulic conductivity and AFP are not high. Another factor to take into account is that mixing and transporting low BD media is simpler than doing the same with high BD media.

Particle-Size Distribution

The most fundamental physical characteristic of a porous media that determines its texture is its particle-size distribution (PSD). Most of its physical properties are significantly influenced by the sizes and relative abundance of the existing particles. In a particle-size analysis, different particle sizes or size increments are isolated, and the abundance of each size is then determined. The substance that makes up the medium solid phase is made up of discrete particles of different sizes and shapes as well as amorphous substances like colloidal organic matter. The nature and behavior of the medium, including its internal geometry and porosity, interactions with fluids and solutes, as well as its compressibility and strength, are largely determined by the particle size and mineral content. The phrase "medium texture" refers to the size range of the particles in the medium, i.e., whether the particles making up a certain medium are mostly large, small, or of some intermediate size, and whether they have a wide or relatively narrow range of sizes. As a result, the phrase has meanings that are both qualitative and quantitative. It characterizes the substance's quality by stating whether it is smooth and fine or coarse and grippy. In terms of numbers, it refers to the proportions of the different size ranges of particles inside a specific medium as well as the carefully observed distribution of particle sizes. Because of this, the medium's texture is a core characteristic and the one that is most frequently used to describe its physical composition. The PSD makes an effort to break down what is typically a continuous range of particle sizes in nature into discrete fractions.

The graphs below displays the PSD curves for two varieties of tuff (scoria) and perlite. The distribution of pore sizes (radii), which will be explored in more detail below, was related to the features of water retention and hydraulic conductivity, and this is how PSD is shaped. The diameter of the largest grains in the assemblage and the grading pattern, or whether the substrate is made up of discrete groups of particles, each of uniform size, or of a more or less continuous array of particle sizes, are two pieces of information that can be obtained from this PSD representation. Poorly graded media are characterized by a step-like distribution curve and a predominance of particles of different sizes. A medium is said to be highly graded if its distribution curve is smooth and flat. It may be inferred from the PSD shapes of tuff and perlite that RTB tuff (volcanic ash, scoria, Merom Golan, Israel) and A13 perlite, whose distribution curves are more step-like, are both weakly graded in comparison to RTM tuff (volcanic ash, scoria, Merom Golan) and B12 perlite, respectively. For calculating soil hydraulic parameters including the water retention curve (RC) and saturated and unsaturated hydraulic conductivities, PSD is utilized in soil science. Particle size and hydraulic characteristics typically have mathematical correlations that are reasonably excellent for sandy soils but less precise for soils with higher clay content (Cornelis et al., 2001). Arya and Paris (1981) made significant advances to the PSD-based prediction of water RCs. The primary foundation of their physico-empirical approach is the resemblance between the cumulative PSD and () curves' forms. Although attempts have been made to find such correlations for soilless medium, numerous authors have created models that are comparable to one another.

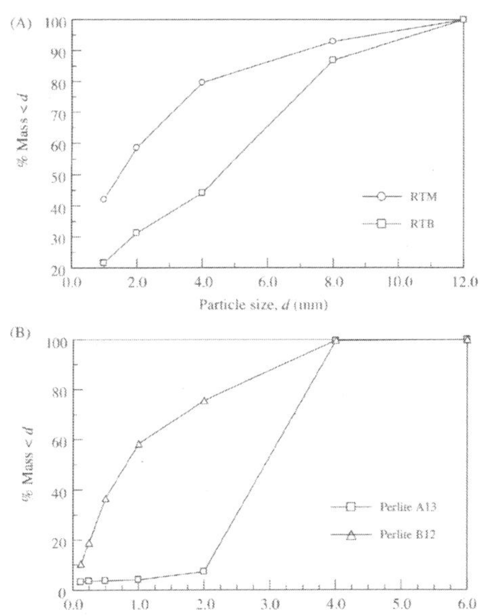

Figure 17: Particle-size distribution for (A) tuff and (B) perlite.

Source: Wallach et al., (1992a); Orozco and Marfa, (1995).

Porosity

TP and its constituent parts are shown as a share of the medium's overall volume. Total pore space, or TP, is the sum of the volumes of the gaseous and aqueous phases of the medium. The order, size, and shape of media particles are all factors that affect TP. Porosity and its components can be determined using the numerous BD measurement techniques. BD frequently has an inverse relationship with TP. However, if materials with closed pores, such perlite or pumice, are employed, the medium BD cannot accurately calculate TP. Non-accessible holes exist in substrates other than mineral ones. Different coir batches were sampled by Evans et al. in 1996. They were surprised to see that the samples showed reduced BD with decreased TP. This, of course, can only occur if a sizeable portion of the pores were not saturated with water during the measurement's saturation stage, indicating that those holes were likely highly hydrophobic. Contrary to closed pores, unsaturation of organic substrates may be considered a flaw in the testing procedure because, in real-world applications, the hydrophobicity of such substrates will be lost and all inaccessible pores will be routinely filled with water. The term "effective pore space" refers to the volumetric amount of water,, required to completely saturate a given volume of a substrate. The volume of closed pores that are not accessible to water is made up of the difference between total and effective pore space. The volumetric proportion of the medium that is air-filled at the conclusion of free (gravitational) drainage is known as the AFP in horticulture. The majority of media and mixes have an AFP of 10%–30%. The size of the container and the frequency of irrigation can have a significant impact on the ideal AFPs because perched water tables continue to sit at the bottom of growth containers after free drainage has ended. AFPs of.20% are required for the rooting of cuttings under intermittent mist. For bedding plants cultivated in shallow trays or plugs, a slightly lower AFP is needed. On the other hand, deep containers with slowly developing, infrequently watered tree saplings may only require an AFP of 10%. It is crucial to take into account that most root systems have a tendency to grow gravitropically and to accumulate into a thick layer at the bottom of the container for all types of containers and media. This could explain why certain plants in media that would normally be regarded as appropriate in terms of AFP perform less than optimally.

Pore Distribution

The structure of a porous medium consists of a 3D network of pores. Large pores are crucial in allowing water, gas, and roots to permeate the medium. Plant roots may access the soil more readily the higher the density of big pores (macropores). Similar to this, gases can interact with the atmosphere more readily the more continuous the macropores are. Continuous macropores directly influence soil solute transport and water infiltration as well. Existing pores limit the ability of roots to penetrate them, but they promote root growth. Numerous studies have demonstrated the enormous advantages of continuous macropores for root growth. In addition to water and nutrient content, soil aeration is one of the most crucial elements affecting soil fertility. The routes for gas exchange between the growth media and the atmosphere are large pores. It makes sense that continuous, big pores help move water. It is now widely accepted that the size and connectivity of soil pores significantly affect how water flows and how solutes are transported through soil. Unless the distributions of the voids, aperture sizes and shapes, depths of penetration, and interconnectedness are known, fluid flow via well-defined structural voids cannot be predicted, according to Jury and Fluhler (1992). The TP and pore-size distribution, tortuosity, and connectedness of a medium dictate its capacity to store water and air as well as its capability to supply them to the plant (via its hydraulic conductivity and rate of gas exchange).

One of the most important 3D factors of pore structure is tortuosity. It is described as the ratio of the shortest distance measured in the direction of the pore to the effective average path in the porous medium. Tortuosity, which indicates the degree of complexity of the sinuous pore path, is a dimensionless factor that is always larger than one. Tortuosity, which indicates greater flow resistance because of the pore system's longer path length, is easily correlated with the conductivity of a porous material. The amount of independent routes that can be taken in the pore space to get from one location to another is known as connectivity. In other words, the amount of nonredundant loops encircled by a particular geometric shape is what is known as connectedness. Every macropore network has connectivity, which is a positive integer corresponding to the variety of closed circuits that may be formed between any two network nodes. The connectivity is equal to 0 if there is just one open circuit and to 1 if the circuit is closed. To reflect the distinct differences in their composition, structure, and stability compared with mineral soils, traditional concepts and values related to physical properties, such as AFP, BD, available water, hydraulic conductivity, gas diffusivity, and field capacity, must be modified for growing media. The absence of standardized nomenclature pertaining to their classification into distinct size ranges is a significant issue with regard to the characterization of soil/soilless-growing media pores. The necessity for a common classification system has been acknowledged by numerous researchers, and ideas for one have been offered. Examples include the suggested micro, meso, and macro-porosity classification by Luxmoore and the proposed index for soil pore-size distribution by Cary and Hayden (1973). (1981). Although various porous materials' pore-size distributions could be determined using a variety of approaches, such as mercury intrusion and water desorption, none of them have been used with soilless container media. Soil physicists are aware that the water RC fundamentally represents the pore-size distribution curve. The capillarity equation is the expression that links pore size to the equivalent suction of water in porous media. Arya and Paris (1981) developed equations for water RCs using pore-size distributions.

Wettability

The existence of soil-water repellency (SWR) has been demonstrated, particularly in field studies where the resulting generation of fingered-shaped flow in the root zone has been observed (Diehl, 2013). However, the impact of a nonzero contact angle (CA) between soil particles and a water drop on soil hydraulic properties and flow, as well as its modeling, have lagged behind. This is true even for soilless surfaces, despite the fact that it was previously known that organic substrates could reject water. The interaction of a liquid and a solid results in wetting. It can involve a liquid covering a surface, dripping into a porous material, or being displaced by another liquid. Wetting is the term used to describe the macroscopic effects of molecular interaction between liquids and solids in close proximity at their interface. The development of a CA at the solid/liquid/fluid interline, the spreading of a liquid across a solid surface (displacing the fluid initially in touch with that surface), and the wicking of a liquid into (and displacement of a fluid from) a porous medium are examples of these manifestations. The dissolving or swelling of the solid by the liquid or chemical reactions between the phases that alter the system composition are not included in the wetting interactions of interest in the current review. The CA, which is defined at a position along the solid/liquid/gas interline as the angle between the normal to the liquid/solid and liquid/gas interfaces at that point, is frequently used as a quantitative indicator of the wetting of a solid by a liquid. This is frequently equivalent to the angle seen in the profile between the liquid/solid and liquid/gas interfaces that are intersecting.

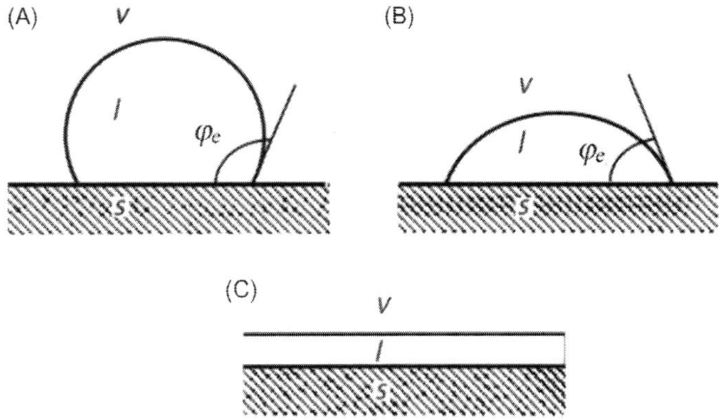

Figure 18: Different wetting situations.

Source: Raviv 1997.

Systems in which the liquid "wets out" the solid (C), partially wets the solid, ϕe > 90 degrees, and 0 degree <ϕe > 90 degrees are those in which the liquid exhibits a CA of 0 degrees against the solid (A and B). The last two situations are identified as surfaces that are, respectively, "water repellent" and "subcritical repellent." Depending on whether the liquid is moving forward or backward across the solid surface, measurements of CAs in real systems show significant variations. One of the following two events typically happens when a liquid drop meets a surface:

- the liquid wets the surface; a circular three-phase contact line is formed that continues to increase in radius and liquid spreads over the entire surface and
- the droplet contacts the surface and the three-phase contact line advances to a certain radius and stops, creating a sessile droplet in the shape of a section of a sphere with a CA that depends on the chemical compositions of the liquid and the surface and the topography of the surface.

Young's equation, which connects the CA to the three interfacial free energies, provides a quantitative description of which of these occurrences occurs and the value of the CA that forms in the second instance.

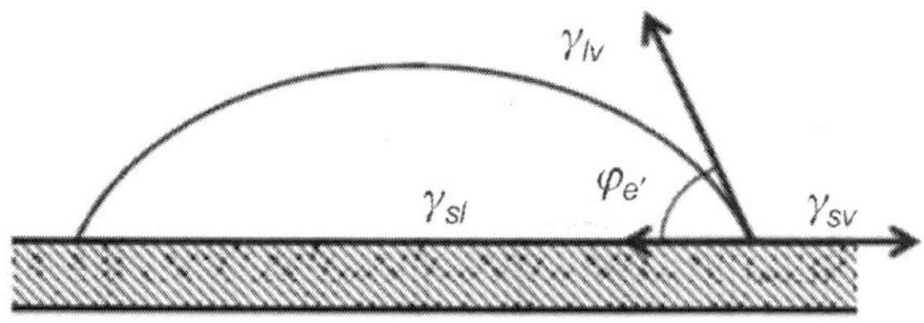

Figure 19: Young's surface tension force.

Source: Young

CHEMICAL PROPERTIES OF SOILLESS MEDIA

Common characteristics of most soilless media include root restriction, a small substrate volume, a low capacity to buffer solution composition, and a finite supply of nutrients. In the horticulture sector, it is widely acknowledged that fertilizing soilless-grown plants requires more effort than fertilizing soil-grown plants. As a result, efficient plant production requires paying particular, distinct attention to the chemical properties of the substrates. Particles and other substances in horticulture substrates have permanent or varying electrical charges on their surfaces. The ionic composition of nutrient solutions is influenced by the surface charge qualities and the dissolving characteristics of the substrate materials. Usually, substrates' chemical characteristics are determined before they are used to grow plants; as a result, undisturbed, homogeneous solid phases are discussed. However, organic substances expelled by plant roots or brought on by breakdown processes build up in the substrate during the growth phase (Raviv, 1997). As a result, the newly formed solids' surfaces become heterogeneous, and the mixture's chemical characteristics may change dramatically from those of the single, well-defined component that was characterized before usage. In comparison to plants grown in soil, plants developing in soilless culture often have smaller root systems. Because the root density of plants cultivated without soil is higher, they have a bigger effect on the rhizosphere than plants grown in soil. Even though there has been a lot of research on managing irrigation and fertilization or controlling plant growth in soilless culture, there is little literature on the chemical characteristics of soilless media. Although the horticulture sector uses a lot of peat, perlite, stone wool, and combinations of organic and inorganic materials, there is not enough information in the literature about the chemical properties of these media. Tuff is a volcanic substance that has undergone extensive investigation into its chemical qualities as a horticultural substrate. It is utilized as a substrate for horticultural crops in Italy, Spain, France, Turkey, and Israel. To illustrate and clarify the chemical processes occurring in soilless media systems, the chapter uses tuff as a model.

Charge Characteristics

In horticultural substrates of volcanic or biological origin, the surfaces of solids hold constant and/or varying positive and negative electrical charges. In layer silicates, isomorphous substitutions—that is, the replacement of structural cations with lower valency cations that have the same coordination number and size—result in a permanent negative charge (common examples: Si41 by Al31 or Al31 by Mg21). The cation exchange capacity (CEC; cmol/kg) measures the degree of cation adsorption to surfaces and is used to describe the medium's cation exchange characteristics. The table below lists the CEC of several organic and inorganic horticulture substrates. It's crucial to keep in mind that plant development may have an impact on the chemical characteristics of the substrate solids when comparing these values to CEC measurements of intact materials taken prior to usage. Typically, the CEC values of soil components are expressed in terms of weight (i.e., cmol/kg). However, because the fixed root zone volume in soilless production methods is so small, it is more practical to connect the CEC of the substrate to the volume unit (cmol/L). It is feasible to appropriately evaluate both "heavy" and "light" material by employing volume-based comparison.

Table 4: Cation Exchange Capacity (CEC) of Some Horticultural Substrates

Substrate	**CEC**	
	cmol/kg	**cmol/L**
Inorganic Substrate		
Perlite	25-35	2-4
Stone wool	34	5
Tuff	10-60	10-60
Clinoptilolite (zeolite)	200-400	400-800
Organic substrates		
Coconut coir	39-60	2-4
Peat	90-140	7-11
Pine bark	98	10
Compost	160-180	15-20

Source: Raviv (1997).

The main source of the variable charge, which can be either positive or negative depending on the solution's pH, is the H+ and OH-adsorption on solid surfaces like metal oxides, hydroxides, microcrystalline silicates (such as allophane and imogolite), or functional organic groups. By measuring the CEC and the anion exchange capacity (AEC) over a range of pH values, it is possible to experimentally evaluate the impacts of solution-pH on the size of the surface charge. The point of zero net charge, also known as the pH at which the AEC and CEC are identical, is widely used to describe the charge characteristics of a medium. Keep in mind that pH is always used to refer to the solutions. A pH titration curve is performed using an indifferent employing an alternate experimental technique. The graphic below shows a typical potentiometric titration for hardwood peat. The point of zero salt effect is the pH level where a series of titration curves for different electrolyte amounts meet. But typical substrates like compost, peat moss, tuff, and stone wool rarely have just one distinct component. They frequently have a combination that exhibits both constant and fluctuating surface charges. This makes it difficult to interpret the experimental CEC, AEC, or pH titration results that were produced from these mixes. Direct interpretation of analytical results is further complicated by the presence of additional sources or sinks for H+/OH- resulting from the dissolution or precipitation of minerals during CEC, AEC, or pH titration measurements.

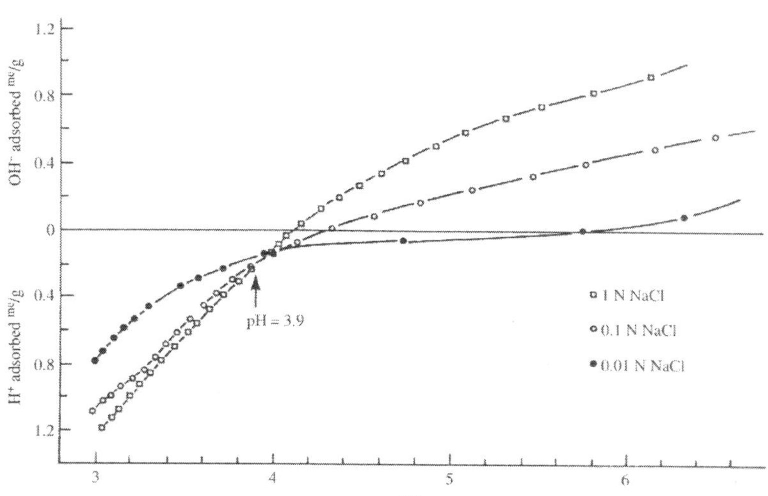

Figure 20: Acid base potentiometric titration curves of woody peat.

Source: Bloom, (1979).

The case of tuff demonstrates the challenges associated with figuring out the true surface charge of a heterogeneous material that has both pH-dependent and permanently charged surfaces, together with primary and secondary minerals. Commonly utilized in greenhouse production, yellow, red, and black tuff are distinguished by their degree of weathering, with yellow tuff being the most weathered and black tuff being the least. For each unit rise in pH, the CECs of yellow, red, and black tuff were found to be pH-dependent, rising by 6.6, 4.2, and 1.9 cmol/kg, respectively. The three types of tuffs' different CEC slope values reflect how much amorphous material with varying charged surfaces they contain. The existence of components with a permanent charge may be indicated by the comparatively high CEC of tuff at pH 3.5, particularly that of the weathered yellow tuff (35 cmol/kg). The tuff kinds' pH titration curves diverged substantially from those of homogeneous variable charged materials. The pH titration lines obtained with three different NaCl concentrations overlapped for the less-weathered tuffs (red tuff and black tuff). Common organic substrates like peat, pine bark, or composts often have high CECs (80–160 cmol/kg) that vary on the pH. The charge is primarily produced by the ionization of COOH groups, with some of it also coming from phenolic OH. Compared to soil inorganic components, the pH effect of CEC on organic material is more pronounced. A unit pH increase adds roughly 51 cmol/kg of organic C to the CEC of soil organic material. The CEC of soil organic matter increased by 29.6 cmol/kg for every unit increase in pH, assuming that organic C made up 58 percent of the total organic matter. The cation exchange capacity (CEC) of pine bark media grows linearly between pH 4 and 7, and for particles smaller than 0.05 mm in size, a unit pH increase contributes 23.1 cmol/kg to the CEC. Composting time affects the charge characteristics of composts, mostly because organic elements, such as the C/N ratio, humic material, and lignin concentrations, change over the composting process. In order to calculate the CEC and maturity indicators of compost quality, the C/N ratio and the lignin concentration were connected with the composting time.

Nutrient Adsorption to Exchange Sites

Ion properties including valence, size, and hydration status have an impact on the affinity of cations for negatively charged surfaces under equal concentration in the aqueous section of the substrate. Due to their higher charge, divalent cations have a stronger affinity than monovalent cations, and typically, the relative adsorption strength follows the order of hydration, with K=NH4> Na and Ca>Mg. The concentrations of each cation, however, are not equal in a typical irrigation solution for greenhouse crops, with the K concentration frequently being higher than the concentrations of Ca, NH4, or Mg. For instance, the concentrations of K, NH4, Ca, and Mg in the nutrient solution are advised to be 5.0, 1.0, 4.5, and 2.5 mmol/L, respectively, for the soilless production of ornamental and vegetable crops in the Netherlands. Therefore, it is anticipated that the exchange reaction, which is the displacement of cations from the substrate solids by K1, will reduce its concentration and then increase the concentration of other cations in the substrate's liquids. The exchange reaction may be advantageous for horticulture when;

- In the liquid solution of the growth media, the concentration of a harmful or toxic cation, such as heavy metals or Na1, decreases due to adsorption or
- Desorption results in an increase in the concentration of an advantageous cation in those liquids.

In contrast, cation exchange could have the reverse effect, with anticipated negative repercussions including include

- increasing levels of a harmful or poisonous cation in the solution after desorption or
- following its adsorption to substrate solids, favorable cation concentrations dropped to a deficient level.

Interactions between Cations/Anions and Substrate Solids

As mentioned above, ion properties like valence, size, and hydration status are related to the solubility and interactions of dietary components with substrate solids. Because NO3- has a very high degree of solubility and has a very low affinity for positively charged surfaces when compared to H2PO4- or SO24-, its concentration in the substrate solution is essentially unaffected by precipitation and/or adsorption reactions. NO3- is the main source of nitrogen for plants grown in soilless culture. There are further details about how bacteria affect the availability of NO3- in the rhizosphere. Similar to NO3-, NH4+ is also a source of nitrogen for plants growing in soilless production, and it typically does not precipitate. Because of its strong attraction for negatively charged surfaces, NH4+ is less available to plants than NO3-. Although potassium is a crucial nutrient, it is not significantly involved in precipitation reactions. K+ may, however, be absorbed by negatively charged surfaces because it is a charged cation, which may limit its availability to plants. It should be noted that the relative concentration of K+ and other cations in the solution and the make-up of the exchange sites determine how they are distributed between the aqueous and solid portions of the substrate. The divalent cations Ca2+ and Mg2+ have a very high affinity for negatively charged surfaces and are heavily involved in precipitation processes, particularly with H2PO4- and SO24-. Actually, it is widely acknowledged that Ca2+ and Mg2+ activities control P availability in the rhizosphere. Metal micronutrients like Cu, Fe, Mn, and Zn are categorized as selectively adsorbed ions (SAIs) because of their extremely high and noticeably greater attraction to charged surfaces than other cations. These metals have a very poor availability to the plant since they are also heavily entangled in precipitation reactions. Zinc and phosphorus were chosen for the following in-depth discussion to demonstrate how cations and anions interact with the substrate's solid components because;

- both are important nutritional elements for plants;
- under common agricultural practices, low availability in the vicinity of the roots of either of these elements often limits crop growth and productivity, even if P and Zn concentrations in the irrigation solutions are adequate;
- ions of both elements are classified as SAIsbecause of their high affinity to charged surfaces; and
- both elements easily form new solid compounds as a result of precipitation reactions.

Phosphorus

Following fertilizer application, there is a common phenomenon of a continuous drop in solution P concentration (PW), which frequently limits crop output. PW declines by means of two mechanisms:

- very fast (on a scale of seconds or minutes) electrostatic reactions of adsorption onto the substrate, as a result of the high affinity of P ions to charged surfaces, and
- slow (on a scale of hours or days) formation of new solid metal-P compounds (with Al, Fe, and Mn under acidic pH conditions and with Ca and Mg under neutral pH or basic pH).

The solution ionic strength (*I*) may affect P solubility via three mechanisms:

- Ion activity coefficient change: An increase in I causes an ion activity coefficient drop, which raises the potential concentration of P in the substrate solution.
- Variation in charge density: Since the diffuse double layer's charge density is influenced by the solution's ionic strength, the impact of I on P adsorption will depend on the pH. Similar to how I affected the surface charge properties.
- Variations in the concentrations of anions and cations in the substrate solution: As the concentration of I rises, anions and cations that may have been adsorbing to or naturally occurring in solid phases may be released.

Phosphorus fertilization management is significantly impacted by the retention of phosphorus by substrates. For instance, 50–60 kg of tuff per m2 is a normal substrate quantity for growing vegetables or flowers. In a desert area, roses (Rosa* hybrida) typically evapotranspire 5 to 10 liters per square meter per day, therefore their daily irrigation requirement is 0.1 to 0.2 liters per kilogram of material. The daily P input is 3–9 mg/kg given that the recommended P concentration in irrigation water for cut rose culture is typically 30-45 mg/L. The probable PW at pH 7.3 with this daily P rate, assuming no leaching, will be less than 0.05 mmol/L, according to the P retention curve. Given the aforementioned information and the presumption that adsorption reactions regulate P availability in tuff suspensions, it is anticipated that P solubility in the solution will decline very quickly, causing the PW to become deficient and insufficient for rose production soon after irrigation.

According to Johanson (1978), a low P application (0.16 mmol/L) reduced the yield of cut roses by 40% compared to a proper P management strategy (1.3 mmol/L). Low P concentrations in tuff substrates could therefore limit plant growth. If a P concentration of 0.5 mmol/L is assumed to be the minimum for cut roses to develop properly, then yellow tuff should adsorb 840 and 600 mg/kg of P at pH 6 and pH 7.3, respectively. With a solution pH of 6 or 7.3, it would take at least 187 or 133 days, respectively, to fill all the available sites for P adsorption on the yellow tuff surface and achieve the desired PW if the daily amount of P added via fertilization is assumed to be 4.5 mg/kg (0.15 L/kg at a P concentration of 30 mg/L). Cut rose development would undoubtedly be hampered during this time. Because the activity products of Ca^{2+}, Mg^{2+}, and $H_2PO_4^-$ are much higher than the Ksp values of known Ca-P and Mg-P minerals, P deficit will limit plant growth when Ca and Mg concentrations are expected to be 1.5-3 mmol/L, as necessary for optimal plant growth. The immediate result of the adsorption and precipitation reactions described above is that, in the pH range where plants grow on tuff substrates, PW may be lower than plant requirements under equilibrium conditions. It's vital to remember that P deficiencies don't only occur in tough media. As long as nutrient concentrations in irrigation water are maintained at a sufficient level for plant growth, precipitation of P with Ca and Mg at neutral and high pH is anticipated to be a general event in all substrates. Given how pH affects adsorption and precipitation reactions, it makes sense to assume that the ideal P availability range should be between pH 6.0 and 6.5. It's commonly known that organic molecules can bind phosphorus. Pine bark media have been reported to retain a high amount of P, while decreased P availability impeded plant growth. A more effective management strategy to increase P availability in the rhizosphere and to enhance P uptake by plants may be to reduce the interval between successive P additions, which entails increasing fertigation frequency and creating nonequilibrium conditions between successive fertigation events. Alternatively, adding humic and fulvic compounds, such as mature compost, to tuff substrates may considerably lessen the P retention by the substrate.

Zinc

SAIs with an affinity for charged surfaces include zinc ions. As pH rises, Zn adsorption rises, exhibiting typical cation behavior. Hoplite [$Zn_3(PO_4)_2.4H_2O$] solubility may be able to regulate Zn solubility. Throughout the pH range of plant growth on many substrates, Zn concentration in the rhizosphere may be lower than plant requirements, similar to P. The impact of solution pH on the adsorption of Zn by perlite and tuff is also well. Since Zn adsorption by substrates is anticipated to follow the CEC values shown in Table 4 above, some substrates frequently employed in soilless culture, such as tuff, zeolite, and compost, have high Zn as well as other micronutrient adsorption capacities, including Fe, Mn, and Cu. On a volume basis, Zn adsorption on fewer active substrates, like perlite and stone wool, is anticipated to be lower. However, as shown for

Zn adsorption onto used perlite, plant growth may change the surface characteristics of their solid phases, which in turn may cause Zn adsorption to be significant even in an "inert" substrate. The utilized perlite had been employed as a sweet pepper growth medium at the Besor Experimental Station in Israel for six months and had been subjected to the usual irrigation and fertilization routine advised by the Israeli Extension Service. The levels of deficiencies: Given how pH affects adsorption and precipitation reactions, it is sense to assume that at acidic pH levels, optimal Zn availability should increase. For the same reasons as were mentioned above for phosphorus, increasing the concentration of the solution metal is ineffective. To maintain adequate availability of micronutrients, chelating compounds like ethylenediaminetetraacetic acid, diethylenetriaminepentaacetic acid, or ethylenediamine di(o-hydroxyphenyl-acetic acid) are frequently utilized (Parker et al, 1995). Additionally, just like with P, it was discovered that upping the frequency of fertigation was a successful way to improve plant uptake of metals.

Silicon (Si) Solution and Addition of P and Zn

The negative or positive surface charge is enhanced by the addition of SAI anion or cation, respectively. Si ions exhibit pH-dependent adsorption characteristics and are a significant component of substrate solids. Due to the dissolving of volcanic glass and allophane that resembles halloysite, the effects of adding SAI anion or cation on solution-Si concentration are interesting. Si concentrations also rose with decreasing pH in all tuff suspensions. When P is added to tuff solutions, Si concentrations rise. P retention is then followed by the displacement of $SiO2$ from aluminosilicates, either because it is displaced from the surface or because the structure of the substrate solids is partially disrupted. Zn addition to yellow tuff suspensions reduced Si concentrations at all the tested pHs, the opposite of what happened when anion addition was used (see figure below). Therefore, it could be interesting to see how P and Zn addition affect Si concentrations.

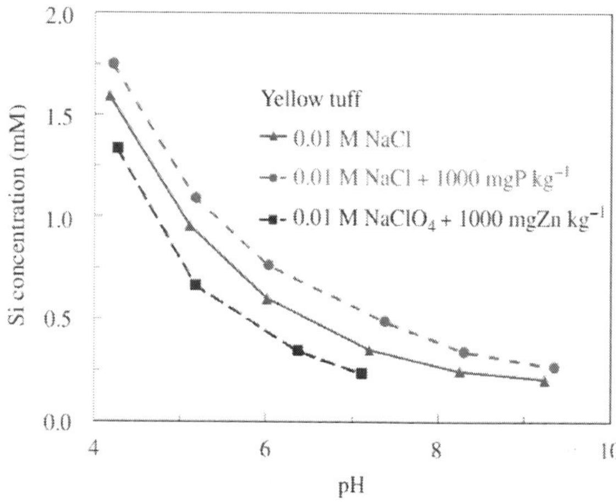

Figure 21: Solution-Si concentrations as a function of phosphorus and zinc addition to tuff suspensions.

Source: Silber, (1991).

Plant growth may affect the chemical properties of the substrate solids through three main mechanisms:

- By using SAIs like orthophosphates in the fertigation nutrition, solids can change their surface charge and chemical makeup.
- development of roots is the primary reason for the addition of new solid materials, specifically organic compounds.
- accumulation of root exudates and decomposition products, particularly humic and fulvic acids that may cause the surface charge to change or the substrate materials to decompose.

As long as the solution-pH is not severe (below pH 4) and does not significantly dissolve the solids, the effect of SAI addition on the chemical characteristics of substrate solids (mechanism 1) is expected to be reversible. Less is known about the direct impact of plant growth on the chemical qualities of the substrate, or the change in chemical properties of the medium brought on by the production of new organic material in the form of roots. The CEC of dry root matter is often in the same range (10–100 cmol/kg) as soil components. The root CEC is pH-dependent, and the same surface complexation model (single capacitance model) that was used to evaluate the adsorption to soil could be used to evaluate the adsorption of protons and copper to root cell walls, indicating that the same mechanism of adsorption is similar to that of soil. However, because of the high CEC values associated with root components such as cell walls, it is unlikely that complete roots contribute to the charge properties of a medium. Because of the poor accessibility of ions from the exterior solution to the interior apoplast, the "real" CEC of an undamaged plant root in vivo is likely lower even though high CEC values may indicate the "potential" CEC. Hinsinger et al. reviewed the impacts of root excretions, root-induced pH changes, or alterations of nutrition availability in the rhizosphere (2003). Silber and Raviv have investigated how plant growth affects the chemical properties of the substrate (1996). Prior to being used in rose cultivation, the surface qualities of red tuff (T0) were assessed, as well as after it had been subjected to fertigation solution in both the absence and presence of plants (Tf and Tp, respectively). Different pH values and ionic strengths of the solutions were used for the extractions. The pH and ionic strength were chosen to reflect the pertinent rhizosphere conditions. The entire, unused red tuff (T0) pH titration lines overlapped, making it impossible to distinguish an ionic strength effect. The basic minerals and unweathered volcanic glass make up the majority of red tuff. Only 19% of it is amorphous, which could be responsible for the pH-dependent charge.

The formation constants between the inorganic ions and root-excreted ligands play a key role in the impacts of plant growth on the solubility of nutritional elements in the rhizosphere. For instance, the production constants (log K0) of citric and oxalic acids with K are 1.2 and 1.1, respectively, whereas the corresponding values for the same two acids with Ca, Mg, and Al are 4.8 and 3.8, 4.7 and 3.4, and 9.9 and 13.0. Soluble complexes of Ca-, Mg-, and Al-organic ligands will therefore predominate in the presence of plants, while K-ligand proportions will be minimal due to the buildup of root-excreted organic acids in the rhizosphere. These variations may be the cause of the discrepancy in how plants affect solution-K and solution-Ca concentrations in tuff suspensions: in the presence of plants, solution-K concentrations decreased while those of solution-Ca and solution-Mg increased in comparison to their concentrations in the same tuff leached with fertilizer solution. The increase in the adsorption of Ca and Mg by tuff surfaces during the growth phase and subsequent displacement by electrolyte ions during the titration operation caused the solution concentrations of Ca and Mg to rise. Due to two distinct, concurrently operating mechanisms, the Ca and Mg adsorption increased:

- higher adsorption of Ca and Mg to inorganic particles due to an increase in the negatively charged surface area, and
- enhancement of negatively charged, adsorbed soluble compounds (like Ca-citrate) on positively charged surfaces.

Because solution-P concentrations are controlled by adsorption/desorption and precipitation/dissolution interactions, which regularly lead them to fall to deficient levels, increasing P availability may be the most significant result of plant development. Therefore, it may be beneficial to alter the surface features in any way that increases P availability. Numerous studies have demonstrated that the rhizospheres alterations by organic acids (mostly citric) released by plant roots significantly increases P mobilization toward the roots (Lambers and Poot, 2003). The increased root density of plants grown in soilless production likely has a bigger impact on P availability than does plant growth in soil.

Nutrient Release from Soilless Substrates

In order to lessen the environmental pollution risks posed by excessive fertilization in soilless production, attempts have been made to control nutrient availability in the rhizosphere and to design substrates that could provide nutrients to horticultural crops during their growing seasons. It may be possible to achieve the goal of decreasing excessive fertilization using two strategies:

- the substrate serves as a slow-release fertilizer by being pre-charged with nutrients that are then released during the growth phase, and
- Native nutrient release from the substrate due to chemical or biological processes, such as N mineralization in organic media or native nutrient dissolution from inorganic components.

In order to pre-charge substrates with nutrients, the material must have high CEC or AEC values. The majority of inorganic substrates, including stone wool, perlite, and pumice, have low values for these traits, indicating that they might not be appropriate for this. However, zeolites have incredibly high CEC values of 200–400 cmol/kg. Zeolites are really utilized as nutrient sources in barren substrates and soils due to their strong affinity for K or NH4 ions. Although the high bulk density (1.92.3 Mg/m3) of zeolites prevents their use as a single-component growth substrate, mixed substrates that may contain zeolites, as well as organic or inorganic materials, are utilized all over the world to grow vegetables and flowers. The entire potential pool for K would be 80-160 g/kg if all the exchange sites of a zeolite substrate were pre-charged with K ions. Since tomato and pepper (Capsicum annum L.) crops only acquire 15-20 g of K on average per plant when grown under intense conditions, 200 g of zeolite will be needed to meet the plants' whole K requirements over the course of a growing season. Therefore, the pre-charging strategy can be seen as possibly successful from a quantitative point of view; yet, issues with the release rate may arise. Basically, the best nutrient release rates must correspond to the plant demand, which is obviously low at the start of the growing season and steadily rises during the various growth phases. However, exchange processes with competing ions (in the case of K+, these include NH4+, Na+, Ca2+, and Mg2+) cause the pre-charged nutrients to be released from the surface sites. As a result, the release rate is influenced by a number of variables, including the concentrations of competing ions in the substrate solution, irrigation dosages, and plant activity. As a result, nutrient availability is not actually regulated and may differ from the plant's needs. By combining phosphate rock with zeolite that has been charged with NH4+ ions, methods (1) and (2) for decreasing excessive nutrient runoff were combined (Pickering et al., 2002). While the NH NH4+ ions provided plants with nitrogen during the growth phase, the primary goal of this mechanism was to cause the dissolution of phosphate rock, either through acidification or through exchange with Ca ions, in order to gradually release phosphate from the zeolite over time. Additionally, zeolites can be employed as slow-release sources of K and micronutrients. Except in a few rare instances where phosphate rock is used, the indigenous nutrients generated from inorganic substrates often contribute very little and are almost nonexistent. Contrarily, organic media, particularly composts, contain significant amounts of nitrogen (N), therefore the creation of composts that might provide N during the growth phase [approach (2)], particularly for crops cultivated organically, may be of interest. Typically, throughout the composting process, some of the initial N content of organic elements is lost. In actuality, the low mineralization rate and low N availability of high-quality mature composts serve as distinguishing characteristics. However, by utilizing a high C/N ratio, bringing the pH down, adjusting the air flow, and lowering the composting temperature, it is feasible to lessen the N loss throughout the composting process.

Substrate Mixtures

Individual substrates are frequently combined in a variety of ratios when they have features that are advantageous and complementary for plant growth and output, such as water holding capacity and aeration efficiency. Due to capillary and adsorptive forces, fine textured field soil and organic components like peat or compost frequently lack the coarse particles needed for sufficient aeration and store moisture very tightly in small holes and on particle surfaces. They are frequently combined with coarser materials like milled bark, sand, or aggregated media to form larger holes that quickly drain after irrigation, which creates ideal rhizosphere conditions that may be adjusted for a particular crop. This improves aeration and water retention characteristics. These mixes frequently have bimodal pore size distributions and water retention characteristics, where the larger pore fraction permits ideal aeration while the smaller pore fraction primarily holds water. Aging frequently causes a shift in the particle- and associated pore-size distributions, which in turn affects the hydraulic and physicochemical mixture properties that must be taken into consideration for irrigation and fertigation management. Zeolite has a remarkable capacity for adsorption, strong cation exchange, and dehydration-rehydration abilities. The majority of the water in perlite is held just superficially and discharged under relatively low tension, making it very inert (low buffering and CEC), which allows for efficient medium drainage and rhizosphere aeration. Diatomite is a siliceous sedimentary rock with a low density that resembles chalk and is soft, friable, earthy, very fine-grained, porous, and light-colored. In essence, diatomite is inert. Due to the complementary qualities of the individual components, a mixture of chemically active substrates, such as zeolite (which has poor water retention capacity), and chemically inert perlite or fine diatomite (which has good air/water retention properties), is expected to yield a beneficial substrate. As a result, adding perlite or diatomite to zeolite enhances its hydraulic qualities. On the other hand, zeolite's strong ion-exchange capacity gives media mixes buffering qualities, which reduces pH and nutrient concentration variations in the root environment. Zeolite, perlite, and diatomite all possess a structural stability that allows for repeated use with different crops. The right quantity of amendment results in environmental conditions that closely match what plants need. Typically, the "optimum" is chosen by growing plants in a variety of mixes through trial and error. This trial-and-error method might be replaced with physicochemical relationships to anticipate mixture behavior from well-characterized constituent qualities, which would greatly speed up the generation of soilless cultures and prevent expensive failures. This assumption serves as the foundation for the following evaluation and discussion of the prevalent physical and chemical features of substrate mixes.

SUBSTRATE MIXTURES PROPERTIES

Physical Properties

Bulk Density: BD is a crucial physical characteristic that influences the aeration, water retention, and root penetrability of substrate combinations. The BD of a mixture can be computed as follows when two components are combined and each contributes weight proportionate to its volume fraction: (Pokorny et al., 1986);

$$D_m = \frac{(V_t C_1)D_1 + (V_t C_2)D_2}{V_t}$$

Where;

- D_m is the predicted BD (g/cm3) of the volume mixture;
- V_t is the total sample volume (cm3);
- C_1 is the volume fraction of component 1;
- C_2 is the volume fraction of component 2;
- D_1 is the BD(g/cm3) of component 1;
- D_2 is the BD (g/cm3) of component2;

The volume fraction of sand in unknown mixtures of sand and bark of the same particle size was predicted using the linear increase in BD associated with an increase in the sand fraction in a mixture (Pokorny and Henny, 1984). The combined volume of two or more substrates with varying PSDs is less than the total volumes of each individual component (see figure below). For instance, when 0.0283 m3 of coarse pine bark and 0.0283 m3 of fine sand are combined, the resulting mixture volume is 0.0425-0.0495 m3 rather than the sum of the individual component volumes of 0.0566 m3. It is typical for the BD of the combination to be greater than the BDs of the individual components and the total and air-filled porosities to be lower as a result. This increases container capacity and the amount of water that is available. Additionally, a significant factor in determining the qualities of the final combination is the moisture level of the different components before combining. When dry components are combined, particles tend to pack more densely, producing a higher mixture BD than when moist components are combined (Pokorny and Henny, 1984).

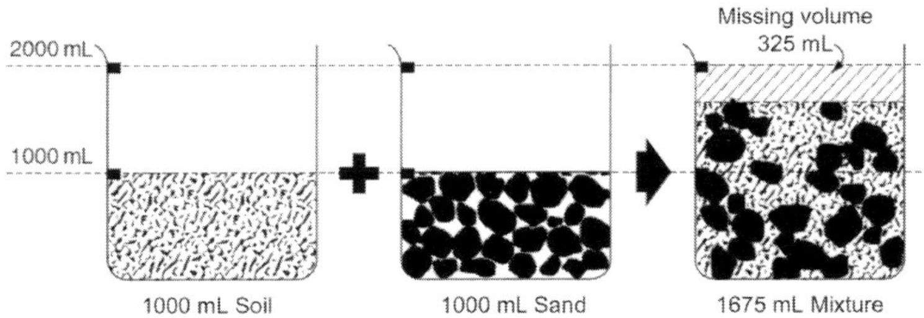

Figure 22: Relationship between individual components and total mixture volumes.

Source: Spomer, (1974).

When coarse bark particles were mixed with fine sand particles, the largest volume reduction in comparison to the sum of the individual component volumes was seen; no volume reduction was seen when medium bark particles were mixed with medium sand particles. The first to show that the effective physical properties of binary substrate mixtures may be anticipated from the qualities of their base components were Pokorny et al. (1986) and Bures et al. (1993). Their theoretical calculations reasonably matched empirically determined mixed parameters (such as BD, total porosity, and water release patterns) for particle-size distribution, particle density, and total porosity. This enables the selection of substrate mixing ratios with the required rhizosphere qualities for certain crops. In order to forecast the BD and porosity of mixtures of spherical particles of different diameters, Bures et al. (1993) used Monte Carlo simulations. The results of simulations for a mixture of sand and pine bark were in good agreement with those of scanning electron microscopy (Bures et al., 1993). Pokorny et al. (1986) devised a different strategy, revising his first equation to take volume shrinkage into account:

$$D_m = \frac{(V_t C_1)D_1 + (V_t C_2)D_2}{V_t - S}$$

S is the volumetric shrinkage fraction.

The equation above can be used to forecast the BD of mixtures made up of three or more components with the adjustments listed below:

$$D_m = \sum_{i=1,n} \frac{(V_t C_i) D_i}{V_t - S}$$

'*i*'is the index of summation and '*n*' the number of components.

Shrinkage S is an unidentified variable in the equation above. A mathematical equation to calculate S has been created based on the shrinkage curves, which may be used to derive this variable (see image below).

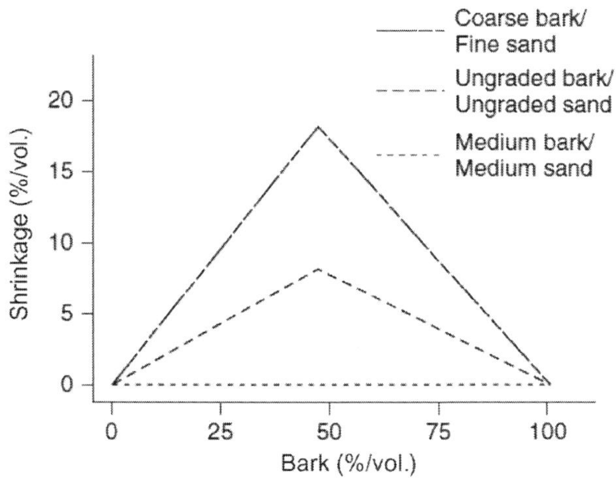

Figure 23: Shrinkage curves for potting media composed of 3 grades of pine bark and sand.

Source: Nash and Pokorny, (1990).

Based Bures and Pokorny (1991) work the following equation can be used to get S.

$$S = 2q_s S_{\max}$$

Where:

q_s is the component with volume <50% and

S_{max} is the maximum shrinkage.

Smax has been demonstrated to correlate with the 1:1 volumetric mixing ratio for a number of binary systems (Pokorny et al., 1986; Nash and Pokorny, 1990; Bures and Pokorny, 1991). For binary mixes of peat moss/perlite, pine bark/sand, and peat moss/sand, the aforementioned equation has been satisfactorily confirmed. So, using the BDs of the individual components and the measured Smax, any binary mixture's BD may be determined. Binary mixes of pine bark and sand produced an excellent agreement between the actual and projected bulk densities, according to Pokorny et al. (1986). The above extended equation for ternary systems is from Bures et al.

$$S = 2(x_2 S_{\max 12} + x_3 S_{\max 13}) \text{ for } x_1 > 0.5$$

$$S = 2(x_1 S_{\max 12} + x_3 S_{\max 23}) \text{ for } x_2 > 0.5$$

$$S = 2(x_1 S_{\max 13} + x_2 S_{\max 23}) \text{ for } x_3 > 0.5$$

$$S = (1 - 2x_1)S_{\max 23} + (1 - 2x_2)S_{\max 13} + (1 - 2x_3)S_{\max 12} \text{ for } x_1, x_2, x_3 < 0.5$$

For a range of volumetric mixing ratios, the aforementioned application has been effectively validated for ternary combinations of pine bark, sand, and calcined clay.

Water Retention:The WRC is a critical substrate property since it affects the aeration and availability of water for developing plants. The substrate pore size distribution and BD have a close relationship with the WRC. The WRC is determined for blends of two or more base materials by the distribution and sizes of pores within each individual component as well as the new pore spaces created as a result of mixing (Spomer, 1974). The sketch below (Spomer, 1974) demonstrates the water retention properties and air-filled porosities of field soil, sand, and a mixture thereof. At matric potentials greater than -330 cm H2O, Sahin et al. (2002) discovered that the water retention characteristics and pore size distributions of a 1:1 volumetric binary mixture of perlite, pumice, and stream sand were highly predicted from the characteristics of the respective base substrates. The perlite/pumice mixture had a better water retention capacity than the perlite/sand and pumice/sand mixes for matric potentials below -330 cm H2O (drier conditions). For the perlite/sand and perlite/pumice combinations, the ratio of macropores (>100 m), which indicate air-filled porosity, and meso- (30-100 m), and micropores (3-30 m), which determine water-holding capacity, was equivalent, but higher for the pumice/sand mixture. In comparison to the other two mixes, the perlite/pumice mixture had a greater fraction of pores >3 m.

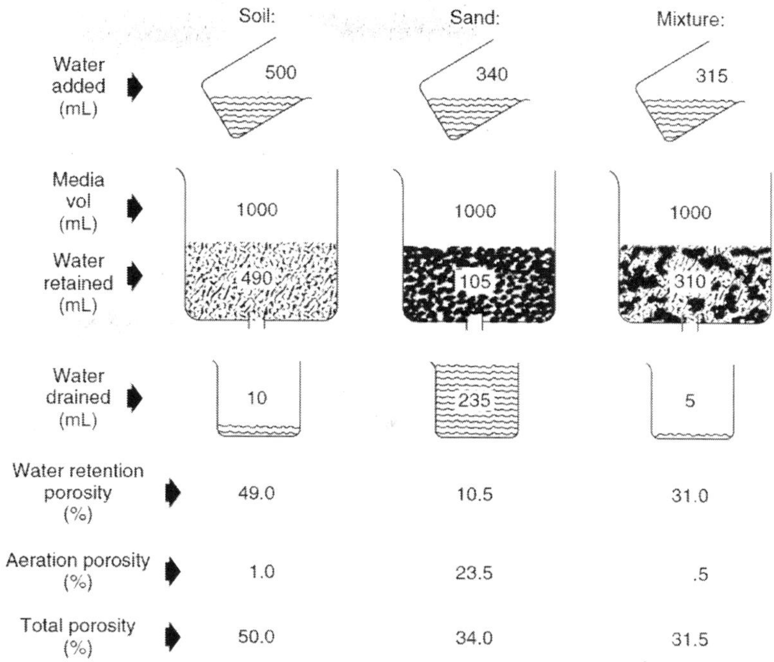

Figure 24: Water retention characteristics and air-filled porosities.

Source: Spomer, (1974).

In comparison to the other two mixes, the perlite/pumice mixture had a greater fraction of pores,3 m. Nash and Pokorny (1992) developed a mixture water retention model that takes volume shrinkage into account:

$$\mathrm{WR}(x) = \frac{\sum_{i=1}^{n} WR(x)_i C_i}{(V_t - S)}$$

Where;

- WR(x) is the water retained within the mixture at matric potential x,
- i is the index of summation,
- n is the number of base substrates,
- $WR(x)_i$ is the retained water in the base substrates at matrix potentialx,
- Ci is the volume fraction (%) of the base substrates,

- Vt is the total mixture volume, and
- S is the volume shrinkage fraction (%).

Although measurements for a variety of pine bark/sand mixtures (0 percent to 100 percent pine bark) at matric potentials of 0, 210, 250, and 2100 cm H2O were reasonably consistent with water retention estimates based on the equation above, this weighting approach has not been independently validated for other substrate mixtures and should therefore be used with caution.

***Hydraulic Conductivity*:** The water flow and nutrient transport through the growth substrates are controlled by the saturated, Ks, and unsaturated, K(h), hydraulic conductivities. Compared to field soils, the fall in K(h) in substrates without soil and decreasing water content is much steeper. A slight decrease in (say, by a few percent) may cause a K(h) value to drop by an order of magnitude or more, significantly reducing the amount of water that is available to plant roots. It has been demonstrated that K(h) and not the matric potential is the primary limiting factor in substrates for water availability and, consequently, for water uptake by the roots (Raviv et al., 2001). For instance, stone wool's K(h) is very low when the environment is less than saturated. The Kozeny-Carman equation can be used to calculate a mixture's saturated hydraulic conductivity, Ks;

$$Ks = \left(\frac{\rho_w g}{\mu}\right)\frac{d_m^2 \eta^3}{180(1-\eta)^2}$$

Where

- ρw is the liquid density (Mg/m3),
- g is the acceleration due to gravity (m/s2),
- μ is the dynamic viscosity (Pa s),
- η is the total porosity (m3/m3), and
- dm is the mean grain diameter (m), which is calculated from either the geometric or harmonic mean as described in Koltermann and Gorelick (1995).

For soils, this technique produces Ks estimations that are often within one order of magnitude of measured values (Koltermann and Gorelick, 1995). The understanding of substrate physical characteristics and our capacity to predict the BD, porosity, and water retention of mixes based on the characteristics of their base constituents and recorded maximal shrinkage have made significant strides. Because most models assume perfect spheres, which are not typical for the naturally uneven form of natural and artificial substrates and mixes thereof, actual measurements of these parameters are nevertheless advised. However, reliable Ks and K predictions are required in order to accept the idea of water availability and its dependency on the transient hydraulic conductivity of the growth substrate (h). Despite the above-discussed advances in the rough estimation of Ks (Koltermann and Gorelick, 1995), mixture Ks must still be determined empirically. The understanding of the correlation between Ks, the WRC, and K(h), however, is important because it avoids the need for the extremely time-consuming and difficult determination of mixture K(h) for matric potential ranges relevant to soilless culture.

Chemical Properties

Stone wool, perlite, pumice, polyurethane foam, and other substrates covered in the previous sections are examples of materials that are chemically inert. In contrast, organic materials, tuff, and zeolite have significant ion-exchange capacities and buffering properties for changes in pH and nutrient availability. While the physicochemical and hydraulic characteristics of many substrates are well understood, chemical characteristics of mixtures—such as CEC, pH, buffering capacity, etc.—are the result of intricate interactions between the constituents of the base combination. The chemical properties of the substrate are anticipated to be proportional to the values of the individual component fractions when a mixture is made up of two substrates that do not react with one another. As a result, the CEC of a milled pine bark/sand mixture declined as the amount of sand increased. Due to the substantial difference in BD between sand and milled pine bark, it was found that CEC could be calculated most precisely on a volume basis. From laboratory examination of each component, the CEC of two-component mixes (pine bark and sand) was predicted (Nash and Pokorny, 1990):

$$\mathrm{CEC}_m = \frac{C_1\mathrm{CEC}_1 + C_2\mathrm{CEC}_2}{C_1 V_t + C_2 V_t - S V_t}$$

The pH rose linearly from 4.1 to 5.4 when the amount of sand in a potting substrate mixture of milled pine bark and sand was increased from 0 percent to 100 percent. The consequences of chemical reactions between components on the aggregate chemical properties, however, have not been modeled in any published works. The majority of research so far has been empirical, offering formulas for altering the properties of chemical substrates with additions. For instance, fly ash is used to change the chemical characteristics of substrate combinations and raise pH. The generated organic acids and microbial enzymes engaged in the breakdown of organic materials may also interact with inorganic materials such as expanded clay granules, vermiculite, pumice, and tuff both chemically and physically (via sorption, complex formation, and dissolution). Through catalytic action, clays are also known to alter a number of OM decomposition reactions abiotically. Clay can therefore have an impact on the chemical composition of substrate mixes. The chemical Characteristics of Soilless Media, the interactions of organic molecules with inorganic minerals, particularly clays, are briefly examined elsewhere.

Common Practice for Mixture Substrate

There isn't much research on the hydraulic conductivity of substrate mixtures, despite the significance of this characteristic for managing soilless growth systems. According to Spomer (1974), the bed depth might be increased or coarse inorganic substrate amendments could be mixed in to facilitate the drainage of organic substrates. Sand is frequently mixed with organic materials, which lowers irrigation water infiltration rates, especially in fresher or less degraded organic materials. In a milled pine bark/sand mixture, the percolation rate decreased as the amount of sand increased. In pure organic substrates, irrigation water quickly percolates to the bottom, but lower penetration rates in sand-amended substrate mixtures allow more complete wetting of substrate particles due to longer contact durations with the infiltrating water. Da Silva (1991) discovered that at the analyzed container height, the hydraulic conductivity of a 1:1 (v:v) mixture of tuff and peat was extremely similar to that of peat, but the WRC of the mixture was slightly closer to the curve of the peat than to that of the tuff (25 cm). The gradual rise in root mass throughout the growth season causes the hydraulic and physical properties of soilless substrates to change, which is a significant concern for the management of these substrates. The displacement of macropore space by successive plant roots is clearly demonstrated by experimental data. In addition to plant roots, microbial development may drastically affect the parameters of the hydraulic substrate. The rate of root and microbiological growth is a crucial consideration when choosing the geometry and size of the container. Due to their great capacity for nitrogen buffering, efficient aeration, and compatibility for a variety of crops, peat moss-based substrates are well-liked. A fine sand or clay is frequently added to substrate combinations that contain light organic components like peat to raise the BD for better mechanical stability. Due to stability issues, very low BD mixes are not recommended for outdoor nurseries.

Sand and clay also make the combinations more wettable. While it is widely known that managing water and nutrients is challenging when the air-filled porosity is low, adding sand or clay has the unintended effect of reducing total and air-filled porosities. Sand, perlite, vermiculite, calcined clay, and vermiculite were among the examined mineral substrates. There was little interaction between total porosity and the PSD of the mineral substrates, and total porosity generally exhibited an inverse relationship with BD. Except that the air-filled porosity values in the fine-grained perlite/peat mixtures were lower than those seen for the individual components and remained mostly consistent for all mixing ratios. In contrast, the addition of coarse-grained mineral substrates increased the mixture's air-filled porosity gradually as the amount of mineral substrates grew; this behavior was noted regardless of the kind of mineral substrate. The addition of fine-grained mineral material had no major impact on the readily accessible water. However, the readily available water was reduced by the coarse-grained mineral particles, especially when they made up more than 25% of the mixture. Vermiculite or perlite addition produced significantly more readily available water than was seen for other minerals.

Similar results were obtained when peat or sawdust was mixed with perlite, pumice, or creek sand at a 1:1 volume ratio. The substrate mixes were made by adding peat moss to the inorganic components at a 1:1 volume ratio. They displayed BD, water retention characteristics, and pore-size distributions that fell between the values observed for the individual components. When coarse- and fine-grained clay substrates (5–25 vol. percent) were combined with peat, the clay component significantly changed the mixtures' water uptake and retention properties. Compared to coarse-grained clays, fine-grained clays caused a more pronounced increase in BD and a decrease in total porosity and readily available water. Even when only a small volume percentage of fine-grained clays were added to the peat, the unsaturated hydraulic conductivity was greatly increased. It is not always possible to ascribe the efficiency of clay supplements to substrate mixtures to specific chemical properties of the clays; physical impacts and interactions with cultivation techniques (such as irrigation frequency and fertilization) are also crucial. The volumetric water content of peat decreases and air space increases when utilized mixed stone wool is added. The volumetric water content of the blended stone wool decreased by 85% when the matric potential was reduced from 0 to 210 kPa, however the water content of peat decreased by only 62 percent when compared to the saturated water content. However, when stone wool is added to the peat substrate, the volumetric water content that is readily accessible is only slightly altered. As long as the fertigation schedule is adjusted in accordance with the physical characteristics of the utilized medium, blended recycled stone wool can be used successfully as a component of peat-based media for geranium production. The physical and hydraulic properties of soilless substrates were improved by the addition of stone wool, creating a conducive rhizosphere environment for crop development and production. When stone wool is added to compost, peat, or other substrates, the resulting mixtures typically have better aeration efficiency and have a higher water-holding capacity.

It is advised to include granulated or powdered clays in substrate combinations to act as a buffer against changes in pH, EC, and nutrient content. The amount of readily available and available water significantly rose as the coir content was raised. Overall, when coir and peat were combined, the mixes' aeration and water-holding capacities showed a nonlinear bulking impact of coir that seemed saturated when coir made up half of the volume of the mixture. The interaction between the hard coir fibers and the flexible peat fibers was thought to be responsible for this effect; coir fibers have the ability to create a framework that is relatively rigid and encourages the open rearranging of peat fibers. Peat, pine bark, and coir are examples of organic elements in mixtures that might be difficult to wet initially and then rewet after drying. To improve the wettability of peat-based substrates, or the capacity of a dry medium to quickly absorb water, clay and mineral substrates, in general, can be modified.

IN SUMMARY

The yield and quality of soilless culture plant production are significantly influenced by the choice of an appropriate growth medium for a certain plant species as well as by proper water and nutrient management. Nutrient ions do not interact with an inert medium, and they lack an adsorption complex. This makes it possible to precisely regulate nutrient levels to correspond to the needs of the plants. Physical characteristics that control plant water availability and aeration effectiveness, as well as offer mechanical support for the plant root system, are even more crucial than chemical characteristics. The initial selection of a substrate that is specifically suited to a given plant is vital because, while the chemical properties of the rhizosphere can be altered during the growing season, it is difficult to change the physical and hydraulic substrate features once a plant has been planted. In order to get the highest yield and highest-quality product, the hydraulic characteristics must be continuously monitored and the irrigation strategy must be modified to account for the dynamic changes brought on by growing root volume and microbial activity. In comparison to field soils, soilless substrates are better for plant health since they are frequently sterile and may be quickly steam-sterilized or fumigated for multiple usages. Although there is a global trend toward growing plants in soilless production systems as opposed to field soils, choosing the best soilless substrate for a given crop is still largely an empirical process that frequently depends on locally available raw materials and economic factors rather than the physical principles that control water, air, and nutrient flow in porous media. The requirement for improved selection criteria for ideal soilless growing substrates has been brought on by recurrent issues with plant hypoxia and water stress. It is also intriguing to note that despite five decades of study, there is still a shortage of knowledge regarding the connections between physicochemical substrate qualities and plant response and performance. To successfully operate soilless systems, the best regulation of water supply and aeration is a crucial requirement.

There are currently no widely established standards for the physical characteristics of substrates used in horticulture. Yeager et al. (1997) listed the following property ranges for "easy-to-manage" substrate mixtures used for commercial horticultural crop production, including irrigation and nutrients: total porosity (50 percent - 85 percent), air-filled porosity (10 percent -30 percent), container capacity (45 percent - 65 percent), available water (25 percent - 35 percent), unavailable water (25 percent - 35 percent), and BD (0.19-0.7 g/cm3). Other researchers question the value of such property ranges because the ideal target qualities depend on the plant type, the growing environment, and the growers' economic goals. The strategy suggested by Jones and Or (1998) appears to have promise. To maximize liquid and gas phase distributions and fluxes to plant roots, they used nonlinear least square's multi-objective optimization, combining the hydraulic features of the media with a control system design. However, only sand and glass bead mixes have been used to evaluate this strategy. More recently, Chamindu Deepagoda et al. (2013) provided a straightforward gas diffusivity mixing model and identified a crucial diffusivity window to help with the choice of base substrate mixing ratios to produce mixes with advantageous aeration and water retention qualities. They also established integral characteristics for a variety of soilless substrates' water, energy, and aeration qualities, which may help in choosing the best growth media and establishing irrigation management set points (Chamindu Deepagoda et al., 2013). Trials in greenhouses are needed to validate these novel ideas. Numerical 3-D modeling of water, gas, and nutrient fluxes is a final approach that may help advance substrate selection prior to greenhouse experiments and provide crucially needed tools for the control and management of containerized soilless production systems. Results from modeling can offer a more robust scientific foundation for the design of commercial greenhouse production experiments and the choice of suitable plant-specific substrates, reducing the likelihood of expensive failures.

CHAPTER 5

MANAGEMENT PRACTICES IN SOILLESS MEDIA

CONTENTS

OVERVIEW

Due to the concentration of soilless crop production in relatively small areas, the release of waste products (such as plastics and synthetic growing media), excessive water use, and use of agrochemicals, soilless crop production has the potential to have an adverse environmental impact. Growers are compelled to use more environmentally friendly cultivation techniques, like closed soilless culture and biological management of pests and diseases, as a result of growing awareness of the pollution connected with soilless farming systems. Closed systems that capture and recycle drainage water use less water and prevent nutrient leaching. These systems do not, however, see much commercial use since managing them is more challenging than managing open (free-drainage) cultivation systems. The greatest challenge for the management of closed systems is the salinity of irrigation water, which is in addition to the potential spread of root-borne diseases. In order to reduce the consumption of water and fertilizers (and subsequently production costs), the environmental impact associated with the disposal of used substrates, and the emission of nutrients and other agrochemicals with drainage water, guidelines for the best management of growing medium and fertigation—i.e., the application of nutrients with irrigation water—are provided in this section.

Growing Systems

All methods for growing plants in solid media other than soil (substrate culture) or in aerated nutrient solution fall under the broad category of hydroponics, also known as "soilless culture" (water culture). The classification of soilless culture takes into account the kind of substrate and container, the method used to deliver the nutrient solution to the plant (such as drip irrigation, subirrigation, flowing, stagnant, or mist culture), and the final destination of the drainage nutrient solution, such as open (free-drain) or closed (recirculating water) systems. The most common soilless cultivation methods are container gardening and aeroponics. Nutrient Film Technique (NFT), floating culture, and other water culture methods are more commonly utilized in research than on a commercial basis, with a few exceptions. For instance, floating culture is increasingly utilized to push bulbs like tulips or to produce fresh-cut green vegetables and herbs. The table below outlines the key features of various hydroponic approaches, including the risk involved with equipment failure and the prevalence of root diseases.

Table 5: Characteristics of various soilless growing systems

	Substrate and drip irrigation	Substrate and subirrigation	NFT	Floating system	Aeroponics
Commercial application	Large	Large	Scarce	Increasing	Rare
Crop type	Fruit veg. Strawberry Cut flowers	Pot plants	Leafy veg-etables	Leafy veg-etables Bulb flow-ers	Vegetables
Substrate	Yes	Yes	No	No	No
Recirculation. Nutrient Sol	Yes/no	Yes	Yes	Stagnant or fairly static	Yes
Investment costs	Moderate/high	High	High	Low	Very high
Running costs	Moderate/high	Moderate/high	Moderate	Low	Fair/high
System's buffer	High	High	Low	High	Very low
Growing risks	Moderate	Moderate	High	Moderate	Very high

Source: Pardossi et al., (2011).

Only crops with quick growing cycles, such green vegetables, could be grown in water. Row crops like strawberry, cut flowers, fruit vegetables (Solanancea, cucurbits), and strawberries are typically grown in containers (substrates) (rose, gerbera, anthurium, etc.). Inorganic or organic substrate, or a blend of two or three distinct materials, such as the peat-perlite or peat-pumice mixture, are utilized in a variety of containers (banquette, pots, bags, and slabs). Drip irrigation up is usually used to provide the crop with a surplus of fertilizer solution (in comparison to the crop's water needs). Subirrigation is being used more frequently when growing ornamentals in pots; the pots are grown in gullies with nutrient solution flowing intermittently through them or on ebb-and-flow benches or floors.

Substrates

Numerous varieties of growth media are utilized in container horticulture. They are often chosen by the producers based on their accessibility, price (including shipping), and local expertise. The growing medium used in substrate culture range in volume from roughly 10 L m-2 (for rockwool or polyurethane slabs) to 40 L m-2 (for perlite bags) (100 to 400 m3 ha-1). Around 40 million m3 (15 million tons) of growth media are used annually in the European Union, including for home gardening. Professional horticulture uses roughly 0.90, 0.14, and 11.9 million m3 of peat, perlite, and mineral wool, respectively. In soilless cultivations, 30.000 to 40.000 m3 of this substrate are replaced annually in Spain. A variety of factors make the final disposal of artificial growth media (or substrate) potentially hazardous to the environment. In reality, especially when they are abandoned illegally, they may contain pesticides and damage the visual amenity of the area. At the conclusion of one or more growing cycles, a variety of substrate types, including mineral wools, are disposed of in landfills; however, landfill costs are rising and many countries no longer have access to them. When Montero et al. (2009) evaluated several greenhouse production systems in Europe using the Life Cycle Assessment (LCA) tool, they came to the conclusion that the fabrication of substrates has a significant environmental impact. Therefore, in addition to reducing the volume of substrate, the reutilization of substrates must be vigorously encouraged. In general, crop production and quality are slightly but significantly reduced when the volume of the substrate is reduced. However, in an experiment done at the University of Pisa with tomatoes grown in perlite bags, no differences were identified between the standard (30 L m-2) and reduced (24 L m-2) bag volumes in terms of crop development and fruit yield.

Open versus Closed System

Drip-irrigated substrate culture may be set up in both open and closed systems. In closed systems, the drainage water is collected and re-used after the pH and nutrient concentration (specifically, the electrical conductivity, EC) are adjusted and, afterwards, disinfected to reduce the risks of root-borne infections. In substrate culture, an excess of fresh (just-prepared) nutrient solution is typically supplied to get over the challenges posed by the uneven transpiration of individual plants and to avoid salt buildup and nutrient solution imbalance. To avoid salinization of the root zone, substrate culture typically uses a drain fraction of at least 20–25 percent. Because of this significant loss of water and nutrients in open soilless systems, operating costs have increased and ground and surface water have been contaminated. According to Malorgio et al. (2001), the yearly drainage loss of water and nitrogen from the open substrate culture of roses was 2123 m3/ha and 1477 kg ha-1, respectively.

Many locations impacted by nitrate contamination have been designated as Nitrate Vulnerable Zones (NVZs) under the European Nitrate Directive (1991). An action plan containing a number of steps is set forth in NVZs to address nitrate loss from agriculture and animal husbandry. Drainage water from soilless cultures that typically has a high nitrate concentration is not at all compatible with the regulations set forth in NVZs. Therefore, the use of closed soilless systems is crucial for protected horticulture to be sustainable. Unfortunately, open soilless cultures are frequently employed for vegetable and ornamental crops since they are much easier to manage. This is in contrast to the Netherlands where the usage of these systems is required. The salinity of irrigation water is the fundamental challenge for the management of closed growth systems, along with the dangers resulting from the potential diffusion of root infections. Ballast ions like sodium and chloride accumulate more or less quickly when the usage of saline water is required. In these circumstances, the nutritional solution is often circulated until electrical conductivity (EC) and/or the concentration of some potential harmful ions reach a maximum permissible value, at which point it is replaced, at least in part ('semi-closed' systems).

SUBSTRATE MANAGEMENT

Substrate Re-use

Reuse is a significant strategy for managing the environmental impact of soilless culture and growing medium in general. Although substrate expenses typically represent a tiny portion of the overall production costs of greenhouse and nursery crops, it may also boost crop profitability. The substrate may undergo a number of changes throughout cultivation that must be considered before it is used again:

- alterations of porosity and water holding capacity due to: organic matter decay, root activity, compaction (fine particle aggregation and segregation), swelling and shrinkage brought on by the wetting and drying cycle;
- changes in pH and salinity (EC); Due to the frequent fertigation with alkaline and/or saline water in the Mediterranean region, greater salinity and pH are typically found at the end of cultivation.
- proliferation of pathogens and parasites (in particular, nematodes);
- accumulation of poisonous substances emitted from the roots or as a result of the decay of organic materials.

After one or two growing cycles, physical alteration of inorganic material hardly ever happens, as was discovered at the University of Pisa using peat-pumice mix or perlite. Depending on the crop type and the substrate's nature, a substrate can be utilized for a certain number of growth cycles. In general, inorganic substrates have a longer lifespan: rockwool up to three years, perlite up to two, and polyurethane up to ten. Due to minimal biostability, organic substrates have a shorter lifespan of up to 2-3 years. Numerous writers examined how crops responded to cultivation in recycled substrates as opposed to brand-new ones. While some of them observed no or only minor variations between virgin and reused substrates, others reported a decline in crop production and/or produce quality in reused media. In general, preventing the spread of pests and diseases is more difficult than managing problems related to excessive salinity and pH. The following actions must be taken in order to extend the use of substrate as much as feasible. Here are some suggestions for reusing substrates:

- Adopt preventative steps to keep infections and pests out of the greenhouse environment;
- To reduce the possibility of disease spreading in the cultivation, clean the recirculating nutrient solution;
- Keep an eye out for infections in the growing environment (including the substrate, fertilizer solution, and plants);
- Before a new outbreak of root diseases and/or pests, disinfect the substrate. if the level is low or moderate
- In contrast, reusing growth media is highly harmful when a crop is seriously afflicted by root diseases or worms, even after disinfection.

The following Prophylactic measures should be observed in substrate re-use;

- Clean and sanitize the greenhouse's structures and floor (e.g., benches, irrigation lines, etc.).
- Avoid bringing in pests and diseases from the outside by wearing clean clothes and shoes and using irrigation water that hasn't been contaminated previously or while being stored.
- Keep crop healthy by:
    - ✓ adequate integration of physical, chemical and biological control tools;
    - ✓ appropriate management of climate and fertigation;
    - ✓ regular scouting and removal of all infected materials;
    - ✓ grow grafted plant, especially if the substrate is re-used.

Substrate Disinfection

For substrate recycling, the employment of physical (steam and solarization) or chemical control (fungicide administration by drip irrigation or fumigation) approaches can be a workable and simple solution. An alternate approach to disinfecting substrates is solarization. On fresh, solarized coconut coir dust that had not been utilized for two consecutive cultivations, Moncada et al. (2008) grew two different tomato cultivars. Regarding crop productivity and produce quality, there were no variations between utilized and virgin perlite. Several investigations showed that the introduction of microbial antagonists that had been preliminarily isolated from suppressive soils and/or used soilless media might induce suppressiveness to root-borne diseases in soilless growth media. In order to create a stable microbial community with the greatest number of beneficial species before the formation of pathogen populations, biological control agents should be administered as soon as feasible. The artificial introduction of particular microbes can also be used in conjunction with other disease prevention strategies like the use of fungicides, minerals (like calcium), or particular watering techniques (e.g., subirrigation coupled with the addition of surfactant to the recirculating nutrient solution).

Nutrient Solution Disinfection

Since the beginning of hydroponics, there has been apprehension that the nutrient solution would act as a vector for pathogens that cause root rot. Recirculating systems have come under scrutiny as a means of spreading the terrible Pythium infection and a whole host of others, despite the fact that they preserve water and nutrients. However, the hydroponic sector has seen a boom in treatments and ways of preventing and controlling such scourges, and growers of all sizes have a variety of alternatives when it comes to solution disinfection techniques. Many of the disinfection techniques typically employed in hydroponics are not as simple as we would like to believe, despite the fact that putting on the newest gadget or dosing up the reservoir may seem like a perfect solution. The crop root zone and the nutrient solution in recirculating systems are finely balanced environments or ecosystems that contain living plant roots, organic matter, a variety of microbial communities, nutrients, moisture, and gases like oxygen needed for root respiration. Beneficial bacteria typically outnumber and outcompete any pathogens that might make their way into the nutrient solution or substrate inside a healthy root zone. In a hydroponic system, this equilibrium is perfect because many helpful bacteria not only have the ability to reduce disease but also may have additional growth-promoting benefits. Root pathogen outbreaks can happen occasionally, though, and in most cases prevention of such problems is significantly more successful than any kind of control technique. A variety of nutrient solution disinfection techniques are available; while the majority are efficient at eliminating pathogens found in the solution itself, they cannot guarantee that a system will remain free of infection because pathogens found inside infected root systems are immune to the effects of treatment. Commercially, large-scale hydroponic farmers who employ biological, UV, ozone, heat, and filtration solution disinfection procedures choose those that provide the least threat to the crop. As there is a potential for toxicity if rates used are too high or the crop is at a sensitive stage, the use of chemical disinfection agents that are dose-responsive is less desired.

***UV — Ultra-Violet Radiation Disinfection*:** Smaller UV units are accessible through the aquarium/fish sector and can be used in hydroponic systems. UV is a plant-safe and effective technique of disinfecting water and nutrient solutions. The majority of anti-microbial activity occurs in the UVc range of 200-280 nm, and plant pathogens are rendered inactive when UV light hits nucleic acid, which absorbs heavily at or near 260 nm. During treatment, a thin layer of nutritional solution is passed through the UV equipment by the UV lamp. The solution must be clear for the process to be successful, therefore pre-filtration is frequently employed to remove any organic matter or other materials first: 100 mJ.cm-2 for the management of pathogenic fungus and 250 mJ.cm-2 for the control of all pathogens, including viruses, are the recommended UV dose rates for disinfecting recirculating systems. When using smaller UV units, it's crucial to have the right-sized device for the amount of solution that needs to be disinfected and to replace the UV bulbs once their useful lives have been reached (UV bulbs will only be effective for a certain number of hours before they need replacement). The disadvantages of UV treatment include the inability of these systems to distinguish between pathogens and non-target or beneficial microorganisms, as well as the potential for UV to affect iron chelates in solution, necessitating the use of the most stable chelate forms, like Fe-EDDHA, and higher iron application rates.

***Ozone*:** Nutrient solutions must be treated with ozone using an on-site ozone generator that releases ozone-enriched air as tiny bubbles into the solution running through a venturi. Over the course of a particular contact period, ozone dissolves from these bubbles into the solution and reacts with organic stuff, including viruses. One of the main advantages of using ozone for water or nutrient disinfection is that any unreacted ozone turns back into oxygen, increasing the solution's dissolved oxygen concentration. This may lead to greater crop growth. Installation of ozone-generating equipment is crucial since off-gassing into the air around crops might harm them. However, iron chelates can be broken down by ozone treatment, and manganese precipitation may result, so these trace elements need constant monitoring. Ozone treatment kills pathogenic microbes and leaves no residue in the nutrient solution that might be hazardous to plants.

***Filtration*:** Once coarser-grade filters have cleaned the solution of bigger organic particles, membrane filtration can be used to eliminate microorganisms. Microfiltration (pore size 100–1000 nm), ultrafiltration (10–100 nm), nanofiltration (1–10 nm), and reverse osmosis (1 nm) are filtration techniques used in hydroponics. The best membrane filtration systems use many filters to remove increasingly smaller particles as the solution passes through. Numerous pathogens have been successfully removed via membrane filtration, but such systems need to be regularly maintained to be effective.

***Slow-Sand Filtration or Biological Disinfection*:** Slow-sand filtration, also known as biofiltration, is a technique for disinfecting water and nutrients that involves gently passing a solution through various layers of aggregate, which are home to vast numbers of beneficial microbial species. Organic material is present in the upper layers of the sand/substrate screens, while microbial species in the lower levels offer biological filtering that aids in pathogen elimination. With this technique, pathogens in the hydroponic system can be further reduced since the nutrient solution released from the filter is enhanced with advantageous microbial species. The biofiltration system's solution flow rates must be accurate and slow enough to allow for disinfection in order for it to be effective.

***Heat*:** Commercial greenhouse farmers employ heat disinfection (pasteurization) of nutrient solutions as one of the most effective ways to get rid of all kinds of plant diseases. To prevent root rot pathogens like Phytophthora and Pythium, most heat treatment systems first pre-filter the solution to remove organic material, then heat it to a temperature of 203°F for a holding time of 30 seconds. Although heat treatment is efficient and safe for crops, there are certain limitations, including the high energy needed to heat solutions to the proper temperature and the time needed to cool them to room temperature before reintroducing them to the hydroponic system.

Chemical Disinfection: Hydrogen peroxide (H2O2) and chlorine are the two most common chemical disinfectants that can be applied to water supplies and hydroponic fertilizer solutions. While both have the problem of potentially harming crops if overdone, both are excellent at killing viruses when used in the proper amount. The dose, temperature, organic loading, and microbiological composition of the fluid being treated all affect how well chlorine works as a disinfectant. The efficiency of chlorine treatment also depends on the pathogen species and the life stage that is being managed. Pythium zoospores were reportedly suppressed by exposure to 2-2.5 mg/l of free chlorine, while other pathogen species required a greater concentration of 14 mg/l, which was phytotoxic to several popular nursery plants. Symptoms of chlorination poisoning include stunting, leaf drop before its time, and necrotic mottling. An unstable oxidizing agent, hydrogen peroxide combines with any sort of organic substance, including pathogens, to produce H2O and an O-radical. The release of oxygen into the nutritional solution is a side effect of utilizing hydrogen peroxide. Root systems may be harmed if H2O2 is used excessively in nutrient solutions that are circulated. Young, delicate plants are especially vulnerable to H2O2 dosage damage; hydroponic lettuce development was shown to be inhibited at rates as low as 8–12 ppm, while Pythium was killed at 50 ppm and Fusarium was controlled at 100 ppm. In order to reduce the potential of plant damage, rates must be carefully studied and depending on the system's organic loading. Both chlorine and hydrogen peroxide react with organic matter in the nutrient solution.

Surfactants: An additional strategy for infection control involves adding surfactants to the recirculating nutrient solution. Within a minute of exposure, non-ionic surfactants can quickly lyse mobile zoospores of diseases including Pythium and Phytophthora. When the disease is already trapped inside plant tissue, surfactants have little lasting effects on plants that are already infected, despite the fact that they can kill vast numbers of mobile zoospores in the nutrient solution. As a result, adding non-ionic surfactants to nutrient solutions is more of a preventative measure than a therapeutic one. Any method for disinfecting nutrient solutions should be used carefully because many of them eliminate species of microorganisms that are helpful in hydroponic systems in addition to plant pathogens. Knowing the benefits and drawbacks of each approach before using it is crucial because some disinfection treatments may have possible drawbacks and improper application of some may even be phytotoxic to plants, especially young delicate crops.

Substrate Monitoring

Making fertigation adjustments and limiting substrate alterations necessitates a regular growing cycle or before its re-use in a succeeding crop media analysis. In most cases, knowing the substrate's pH and EC is enough to check for anomalies. On the other hand, to learn more about the concentration of certain nutritional or ballast ions, such as Na and/or Cl, the aqueous solution obtained from the substrate can be examined by outside laboratories or on the farm using a fast test. In order to standardize the moisture level at sampling, the NS maintained by pre-shaped substrate (i.e., rockwool or polyurethane slabs) is commonly sampled using a syringe after an irrigation session. It is expected that substrate size matches the container capacity after watering (with adequate drainage). The monitoring of loose substrates, such as perlite, pumice, and peat, involves creating an aqueous extract. An aqueous extract is created by mixing distilled or deionized water at a ratio of 1:1.5, 1:2, or 1:5 with the required volume of substrate (previously moistened to container capacity). While different ratios are utilized for inorganic substrates, the 1:1.5 dilution is employed for organic material (such as peat, compost, and coconut fiber). The biggest disadvantage of the non-destructive pour-through approach, which is another option, is how challenging it is to interpret the results. This technique involves applying a specific amount of water to the pot surface after irrigation in order to force the substrate solution out of the container's bottom and into a collection saucer. A technique based on the Rhizon-Soil-Moisture-Sampler (RSMS) is employed for slabs.

Figure 25: Rhizon Soil Moisture sampler deployed in situ.

Source: Dale, (2018).

The RSMS comprises of a porous plastic extraction tube that is introduced into the substrate and attached to the sample vial via a hypodermic needle. Through a hypodermic needle, a second tube is likewise connected to the vial. On the other hand, the second tube is attached to a hand-operated vacuum pump. The substrate solution is drawn into the extraction tube and subsequently into the vial by the vacuum.

Table 6: Guide values for growing media

Parameter	Extraction method (substrate: water in v: v)		
	1:1.5	1:2	1:5
pH	5.5 - 6.0	5.5 - 6.0	5.5 - 6.0
EC (dS m^{-1})	0.6 – 1.5	< 1.5	0.2 – 0.5
NO_{3-} (mg L^{-1})	40 - 80	50 - 70	10 - 20
NH_{4+} (mg L^{-1})	25 - 35	3-6	8 - 10
K (mg L^{-1})	12 - 45	50 - 100	11 - 16
P (mg L^{-1})	20 - 30	3 - 5	6 - 8
Ca (mg L^{-1})	40 - 80	50 - 80	10 - 20
Mg (mg L^{-1})	25 -45	20 - 30	6 - 10
Na (mg L^{-1})	40 - 60	< 90	10 - 16
SO_{4-} (mg L^{-1})	115 - 150	40 - 90	35 - 45
Cl (mg L^{-1})	60 - 100	< 90	18 - 30
Fe (mg L^{-1})	01 – 0.4	0.5 - 1	0.1 – 0.5
Mn (mg L^{-1})	0.01 – 0.3	0.2 - 0.4	0.01 – 0.1
Cu (mg L^{-1})	0.01 – 0.06	0.05 - 0.1	0.01 – 0.03
Zn (mg L^{-1})	0.01 – 0.3	0.1 - 0.2	0.01 – 0.1
B (mg L^{-1})	0.01 – 0.3	0.2 - 0.4	0.01 – 0.1

Source: Pardossi et al., (2011).

Below are some soilless substrate tests to enable monitoring;

***Step 1: substrate moistening*:** A sample of substrate measuring around 300 ml is placed in an appropriate-sized plastic container, which is then moistened to container capacity by slowly adding distilled water and gently swirling with a spoon or spatula until there is minimal water on the bottom.

***Step 2: Extraction (1:2 dilution)*:** In a graduating plastic bottle, add 400 ml of distilled water first, then the substrate until a volume of 600 ml is obtained. Pour the slurry through a tiny filter funnel lined with coarse filter paper to filter the sample after shaking the bottle for about 2 minutes and allowing the suspension to equilibrate for 15 minutes.

Step 3: measuring EC and pH: For testing substrates, the fundamental criteria are electrical conductivity and pH. With a handheld EC and/or pH meter, they can measure with ease. Remember to regularly (i.e., weekly) verify the calibration of your EC/pH meter.

Step 4: measuring ion content: On-farm assays can quickly identify the concentration of a specific ion in aqueous extracts. When the solution is applied to a reactive strip, a purple color that intensifies in proportion to the nitrate level appears. After a test-specific amount of time, the strip is read (f.i. 60 sec., for nitrate determination).

Step 5: interpreting test results: Check the results against your own reference standards as well as the guiding values created for each extraction procedure. Remember that storing data for each crop is essential for accurate interpretation of laboratory and fast analysis results as well as for early substrate anomaly detection.

Quick Analysis Monitoring

Quick test kits (QTK) are an acceptable substitute for the pricey and time-consuming laboratory analysis of substrate extracts and nutrient solutions. Commercially, QTK is offered in a wide range. QTK is offered by numerous vendors for a large variety of analytes (e.g.: Sigma-Aldrich, Milano, Italy; Merck, Darmstadt, Germany). Ready-to-use reagents, containers, and a portable measuring apparatus often make up QTK. They can be purchased for as little as a few euros (for colorimetric assays using reactive strips) to as much as a few hundred euros (for ion selective electrodes) (ISE). The following types of commercially available QTK are generally used:

- **Titrimetric QTK:** They are based on the reaction of the analyte with a known concentration of reagent; the reaction often takes place in the presence of a material that changes color when the reaction is finished.
- **Colorimetric QTK:** They involve reactions that produce color. The reactive strips are the simplest colorimetric assays. Using a photometer or a reflectometer, one can measure the color intensity of the final compound, which is proportional to the analyte concentration, or one can do so semi-quantitatively by contrasting the sample color with a reference range.

According to research done at the University of Pisa, the following ions could be examined using titrimetric and reflectometric methods in irrigation water, substrate extract, and recirculating nutrient solution:

- Ammonium (reflectometry; 0.2 – 0.7 mg L-1)
- Boron (titrimetric assay; >0.2 mg L-1)
- Chloride (reflectometry; 2 – 50 or 50 – 1000 mg L-1)
- Nitrate (reflectometry; 2 - 90 or 5 - 225 mg L-1)
- Phosphate (reflectometry; 0.1 – 5 or 5 - 120 mg L-1)

A portable multi-ion meter is currently on the market and costs several thousand euros (Clean Grow, Cork, Ireland; www.cleangrow.com). It has a single probe with six ISEs combined. In addition to pH and other analytes, the following ions can be examined: ammonium, calcium, chloride, copper, nitrate, and potassium. There hasn't been any independent testing of this gadget for horticulture use.

Detecting Pathogens and Phytotoxic Compounds

The onset of root disorders can be avoided with an early diagnosis. Pathogens may be found and identified using DNA technology in a variety of crop samples, such as irrigation water, fertilizer solution, plant tissues, and growing medium. DNA tests are sensitive, only the target pathogen is detected, rapid (results may be available in a few days after collection), specific (only the amount of infection is indicated), and sensitive. Regular tests are required to determine whether or not the recirculating nutrient solution has to be disinfected (once a month to once every six weeks). A buildup of plant and microbial metabolites in the substrate of the growth media and/or the recirculating nutrient solution is another issue that could lower crop productivity. While some of these substances may promote the growth of plants, others may be phytotoxic due to their allelopathic nature.

Table 7: Drawback on soilless substrate re-use and their remedies

Drawbacks	Remedies
Lower air capacity Higher water capacity	Substrate re-mixing Adjustment of fertigation regime
Higher pH Higher salinity (EC) Accumulation of toxic ions Nutrient imbalance	Leaching irrigation Adjustment of fertigation regime

Pest and pathogen contamination Allelopathy	Disinfection Use of resistant genotypes Grafting Biological control Suppressiveness

Source: Pardossi et al., (2011).

IRRIGATION

Irrigation Scheduling

The over-irrigation of protected crops leads to water loss and contamination from fertilizer leaching. For short-cycle, soil-grown crops like leafy vegetables, the annual irrigation water demand in protected farming systems ranges from 150 to 200 mm (Lm-2) to 1,000 to 1,500 mm for soilless row crops like Solanaceae and cucurbits. Inadequate irrigation scheduling, which entails figuring out the best irrigation dose and frequency, frequently leads to over-irrigation.

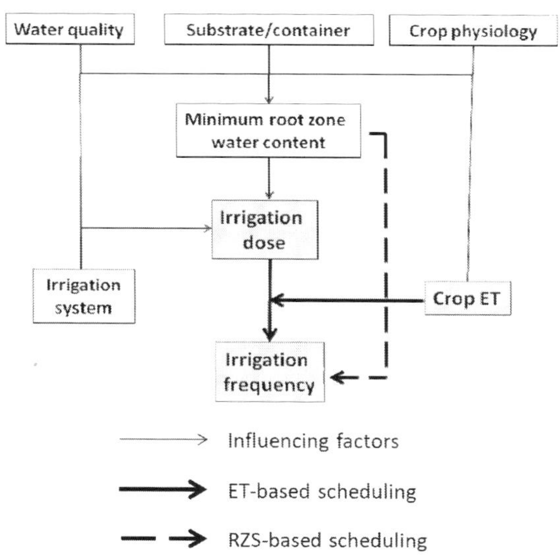

Figure 26: Conceptual framework of irrigation scheduling in soilless culture.

Source: Pardossi et al., (2011).

Crop physiology (i.e., response to water status of the growing medium), water quality (i.e., salinity and concentration of specific ions), and irrigation system layout (i.e., the discharge rate of emitters, minimum irrigation length, irrigation efficiency and uniformity, etc.) must all be understood in order to determine irrigation dose, whereas crop water uptake (WU), which is predominately determined by ET (>90 percent) by ET, determines the frequency. In open substrate, precise irrigation planning is essential because it affects both the seasonal water use (W) and the pollution brought on by pesticide leaching. Conversely, excessive or insufficient irrigation can have an impact on crop growth and yield in closed systems by, for example, increasing the prevalence of physiological problems (such as blossom-end rot in tomato and pepper plants) or the susceptibility to root diseases.

Irrigation Dose

Calculating two quantities—net (I_{net}) and gross or actual (I_{gross}) irrigation volume—is necessary to determine the ideal watering irrigation dose (expressed as L m-2). The first quantity is calculated as: The first quantity corresponds to the maximum oscillation in a substrate that the crop can endure.

$$I_{net} = \frac{AW_{cont}}{100} \cdot V_{cont} \cdot \mathrm{f}$$

Where:

- AW_{cont} (%) is the available water in the container-substrate system under consideration,
- V_{cont} is substrate volume (L m^{-2}) and
- f (dimensionless) is a crop-specific irrigation coefficient ranging from 0.05 to 1.0: the lower f, the smaller the variation in substrate moisture and the shorter the irrigation.

The hydraulic characteristics of the substrate, the irrigation system structure, which impacts water distribution uniformity and establishes the minimum length of each irrigation event, and crop physiology must all be taken into consideration while choosing f. The actual irrigation dose is typically larger than I_{net} as an excess of water is typically required because of:

- the unequal transpiration of individual plants;
- ii) the differences in water discharge of individual trickle nozzles and the consequent uneven water distribution;
- the need to prevent salt accumulation in the root environment.

Hence, I_{net} is calculated as;

$$I_{gros} = I_{net} \cdot K_S$$

Where: K_S (dimensionless) is a safety coefficient.

The percent difference between water supply and drainage for each irrigation event is known as the drain fraction (DF), and it roughly equals:

$$LF = 100 \frac{(I_{gross} - I_{net})}{I_{gross}}$$

Crop and irrigation consistency, as well as the danger of substrate salinization, are all factors in K_S. The safety coefficient ranges from 1.15 (uniform crop and water distribution; use of irrigation water with relatively low salinity; high crop salinity tolerance) to 2.0 (large inter-plant variability in ET; poor irrigation uniformity; use of saline water; salt-sensitive crop); these values produce DF of 0.13 and 0.50, respectively. Most of the time, a K_S of 1.30 (LF = 23 percent) is appropriate. Evidently, K_S determination in closed systems is less important, even though excessive K_S and DF raise the cost of water pumping and NS disinfection. In drip-irrigated substrate culture, the duration (DI) of each irrigation event is typically relatively brief (in the range of a few seconds to a few minutes) and is dependent on the density (d, number of drippers per square meter) and discharge rate (r, L h-1) of emitters. It is calculated as follows:

$$D_I = \frac{d \cdot 3600}{r}$$

Where: D_I is expressed in seconds.

Irrigation Frequency

In general, crop ET divided by Inet is used to calculate irrigation frequency (not Igross). For instance, the number of irrigation events in a day is the outcome of expressing ET on a daily basis. In substrate culture, the crop is often irrigated several times throughout the day, with the initial irrigation taking place in the early morning. The light phase, which lasts 10 hours or fewer in the fall and winter and 12 to 14 hours in the spring and summer, is when more than 90% of daily crop ET occurs. In dry locations or heated greenhouses, watering may also be required in the dead of night. Due to frequent watering, soilless culture irrigation is typically controlled automatically by:

- timer, on the basis of the grower's estimate of crop ET;
- weather station or simple light sensor, on the basis of crop ET predicted with more or less complicated equations (modeling approach);
- weighing gutter (or similar devices), which measures gravimetrically ET (and possibly growth) of a few test plants over a short time (minutes to hours).
- root zone sensor(s); which measures directly θ or ψm.

Timer: As a general rule, distribute irrigation events over the entire 24 hours when irrigation is timer-controlled, as shown in the picture below, which illustrates how daily ET is divided into various time periods (early morning, sunny hours, late afternoon and night).

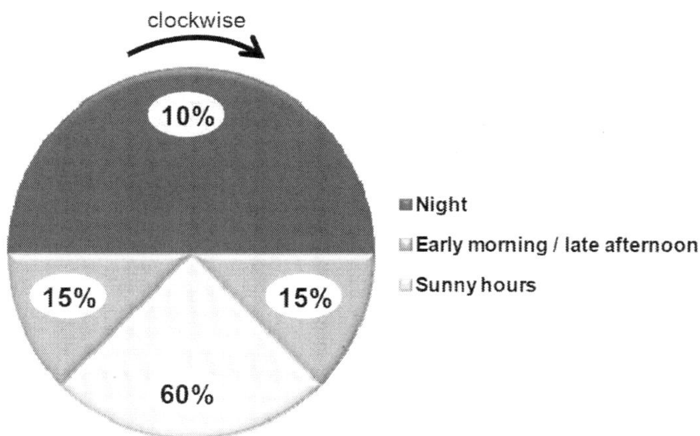

ET modeling: For predicting ET in greenhouse crops, various models of varying complexity have been devised. The FAO PenmanMonteith (PM) equation, one of several methods for calculating ET, is presently regarded as a standard reference (Allen et al., 1998). However, because it involves an understanding of numerous variables and parameters, its implementation is not simple. In order to forecast ET as a function of LAI, intercepted radiation (Rint, MJ m-2 h-1), and VPD (kPa), the following authors developed simpler equations:

$$ET = A \cdot \frac{R_{int}}{\lambda} + B \cdot LAI \cdot VPD$$

Where:

- A (dimensionless) and B (kg m^{-2} h^{-1} kPa^{-1}) are empirical coefficients:
- R_{int}, is estimated from LAI and a crop-specific light interception coefficient (k):

$$R_{int} = 1 - \exp^{-k \cdot LAI}$$

Following the necessary calibration, the equation previously described correctly predicted ET in a variety of greenhouse crops. De Graaf (1988) presented the following straightforward methodology for calculating the daily ET of row crops in heated greenhouses:

$$ET = \frac{h}{m} \cdot \left[a \cdot R + b \cdot \Sigma_{1440}^{1} \min_i (T_t - T_a)\right]$$

Where:

- h and m are actual and maximum heights of the crop, respectively;
- R is the inside global radiation;
- a and b are crop-specific coefficients, the latter being attributable to heating;
- min_i is the successive minutes during the day that the temperature of heating pipeline (Tt) is different from air temperature (Ta). The value of b is 0.22 10-4 for tomato and cucumber grown in Dutch growing conditions.

Simple linear regression of ET versus outside or inside solar radiation has also been proposed for effective management of irrigation in greenhouse crops, as solar radiation is the key climate variable impacting ET in protected crops, especially under unheated greenhouse circumstances. The covering material's light transmission coefficient, which normally ranges from 0.60

to 0.70, must be known in the first scenario. According to expectations, the ratio was higher for crops with higher LAI and ranged from 0.65 to 0.80.

Root zone sensors: By continuously observing or m of the growing media, soil moisture sensors could be utilized to control the frequency of watering and possibly the water dose (Pardossi et al., 2009). While soil and plant scientists have access to expensive and complex SMSs like the neutron probe and TDR (time-domain reflectometry) equipment, commercial crops need irrigation control using inexpensive and useful devices. New SMS types that measure soil dielectric characteristics have created some intriguing opportunities (Pardossi et al., 2009). In comparison to conventional water-filled tensiometers, new dielectric SMSs are less expensive, require less maintenance, and require less user expertise. Numerous studies have proven the use of SMS technology for irrigation management in soil and soilless culture, and at the moment, a variety of straightforward irrigation controllers are offered on the market and interfaced to one or more SMSs. The crop species and growing medium will determine the threshold values for or m. The typical range for m in substrate growing systems is between -4 and -10 kPa (Pardossi et al., 2009). With WRC, this value can be changed to θ. For instance, in perlite, an m of -5 kPa equates to an ψ of 34%. Additionally, sensors have been created for simultaneous measurements of temperature, θ and pore water EC in soil or soilless media, such as 5TE (Decagon Devices) and WET (Delta-T Device) (Pardossi et al., 2009). The possibility of regulated fertigation is made possible by these sensors. A recent automated fertigation system that simultaneously measures the EC and pore water EC (ECPW) of the growing medium using the WET sensor has been successfully constructed and tested to control irrigation frequency and EC of fertigation water. In order to avoid salt buildup in the substrate and reduce water drainage, specific algorithms were implemented in the control software to activate irrigation when a pre-set threshold was reached and to modulate irrigation dose and/or nutrient solution EC (also by mixing different sources of water like recirculated water, groundwater, rainwater, etc.).

NUTRIENT MANAGEMENT

Plant Mineral Nutrition in Soilless Culture

If an artificial substrate is used, it should give the best conditions for crop growth and yield, and a well-balanced and highly-concentrated NS should be available. High concentration is employed in order to:

- fertigation mixer systems that typically dilute 100- to 200-fold concentrated stock solutions with raw water on the basis of the measurement of electrical conductivity can automatically prepare the nutrient solution in order to ensure adequate nutrient supply without the challenges of maintaining a fairly constant ion concentration in the root zone (EC). Low-concentration culture solutions could not be used in industrial greenhouses due to the salt dissolved in raw water and the precision of existing mixers.
- to enhance the quality of the fruit at least in some crops, such as vegetables and fruit (osmotic effects). For high-quality requirements in these cultivations, EC greater than 2.5 to 3.0 mS cm-1 is typically required.

Nutrient Solution

All of the macronutrients—nitrogen, phosphorus, potassium, calcium, magnesium, and sulfur—as well as all of the trace minerals—iron, boron, copper, manganese, zinc, and molybdenum—are present in NS in concentrations on the order of milli- and micro-moles per liter, respectively. pH levels between 5.5 and 6.5 are ideal. Total molar concentrations range between 20 and 40 mM, or between 1 and 2 g l-1, depending on a variety of variables, including crop traits (such as salinity tolerance) and stage, climate, and hydroponic system. Nitrate is the predominant ion on a molar basis in the NS. In general, horticultural crops cannot withstand the ammonium form of nitrogen, thus hydroponic culture does not utilize it or only uses it in very small amounts. The concentration is typically given as EC (dS m-1). Assuming that the concentration of cations is equal to the concentration of anions, a straightforward linear relationship may be utilized to convert equivalent concentration of cations (C+, meq L-1) in EC for the majority of balanced NSs in the range of 1.0 to 4.0 dS m-1 (Sonneveld et al., 1999):

$$EC = 0.19 + 0.095\ C^{+}$$

Through the use of portable tools, fast test kits, and/or laboratory analysis, one can determine the chemical properties of the NS. In closed systems, especially, pH and EC are often and routinely evaluated. Acids, raw water, and stock nutrition solutions are combined to create NS. The use of fertilizers with a high purity grade and solubility is advised. The fertigation system, which typically uses volumetric or electronic injectors, carries out this function. The concentration of stock NSs is typically 50 to 200 times higher than that of NS supplied to the crop. At least two stock NSs are created to separate calcium salts from phosphates and sulfates in order to prevent precipitations of salts, such as calcium phosphate. One or more ions that are frequently present in high concentration in water but are not necessary for plant growth (ballast ions). It is essential to understand the raw water's quality in order to:

- check whether water can be utilized as such or needs specific treatments;
- to calculate the amount of fertilizers required to prepare nutrient stocks.

The table below lists irrigation water standard values. The precise quantity of salts needed to create stock NSs can be calculated using a number of computer applications and spreadsheets that are also available online. Excel spreadsheet is one illustration (NS Calculator).

Table 8: Main water parameters required for fertigation use in soilless culture

Parameter	Units	Degree of restriction on use		
		None	Slight to Moderate	Severe
EC	dS m^{-1}	0-0.75	0.75-2.25	>2.25
Bicarbonates	mol m^{-3} (ppm)	0-2 (0-120)	2-6 (120-360)	>6 (>360)
Nitrates	mol m^{-3}	<0.5	0.5-2	>2
Ammonium	mol m^{-3}	≈0 0.1	-1	>1
Phosphorus	mol m^{-3}	<0.3	0.3-1	>1
Phosphorus	mol m^{-3}	<0.3	0.3-1	>1
Calcium	mol m^{-3}	<1.5	1.5-5	>5
Magnesium	mol m^{-3}	<0.7	0.75-2	>2
Sodium	mol m^{-3}	<3	3-10	>10
Chloride	mol m^{-3}	<3	3- 10	>10
Sulfates	mol m^{-3}	<2	2-4	>4
Iron	mmol m^{-3}			>90
Boron	mmol m^{-3}	30	30-100	>100
Cupper	mmol m^{-3}			>15
Zinc	mmol m^{-3}			>30
Manganese	mmol m^{-3}			>10

Source: Pardossi et al., (2011).

Fertigation System

A fertigation system may be composed by:

- pressured and filtered raw water from single or multiple sources such as surface water, groundwater, rainwater, reclaimed wastewater and, in closed system, recirculating NS (disinfected or not);
- dosing machine (many devices are available on the market; Seligman, 2011), which injects two or more stock solutions into raw water and adjusts the pH of the supply NS by means of acids or bases;
- probes for monitoring EC and pH of supply, recirculating and drainage NS;
- water meters to control the injection of nutrient stocks and water flows in the system;
- climate sensors, which are used to estimate crop ET;
- control system (computer program), which schedules irrigation and adjusts the chemical characteristics of the NS according to the grower's instructions;
- hydraulic components such as pipelines, drippers, electro-valves and tanks. A few of these components (e.g., 5 and 6) are not essential and are installed only in a sophisticated fertigation system.

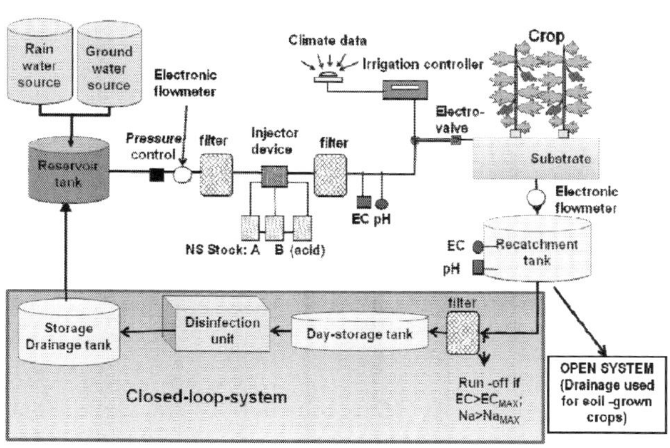

Figure 27: Typical layout of substrate soilless growing system.

Source: Pardossi et al., (2011).

Fertigation Strategy

The criterion used to differentiate between open and closed systems is what happens to drainage water. Although drainage water is not utilized again on the same culture in open systems, it may be used for crops growing in the soil in the field whether the excess NS given to the substrate is collected or not. In closed systems, drainage is gathered, filtered, and pH and nutrient content (specifically, EC) adjusted before being re-used to irrigate the same crop. It is suggested that recirculating NS be disinfected due to the possibility of microbial contamination (see the previous section). In closed systems, evaporation from the substrate and uncontrolled seepage are typically insignificant, therefore water requirements match actual crop water uptake (WU, or ET). If salt water is utilized, ballast ions, which are dissolved in the water at a concentration higher than crop absorption concentration, accumulate more or less quickly (CU, the ratio between the ions and the water taken up by the plants). In such circumstances, NS is circulated until its EC (ECNS) and/or the concentration of certain ions (such as Na, Cl, or trace elements such as B) reach the maximum permissible thresholds value (ECMAX or CMAX) for the crop in question; following this, NS is replaced, at least in part (flushing). In commercial closed systems, three distinct approaches that differ in the criterion for nutrient replenishment could be applied.

Strategy A:

In order to utilize all of the drainage water, the reservoir tank is refilled with refill NS, which is made by combining raw water and drainage NS at a ratio typically equal to the drain fraction (DF) and adding nutritional stocks to achieve a desired EC value. If one or more ballast ions are dissolved in raw water, this technique results in progressive nutrient depletion while maintaining the EC of recirculating NS at a constant level. Evidently, EC in this system does not provide data on the concentrations of certain ions.

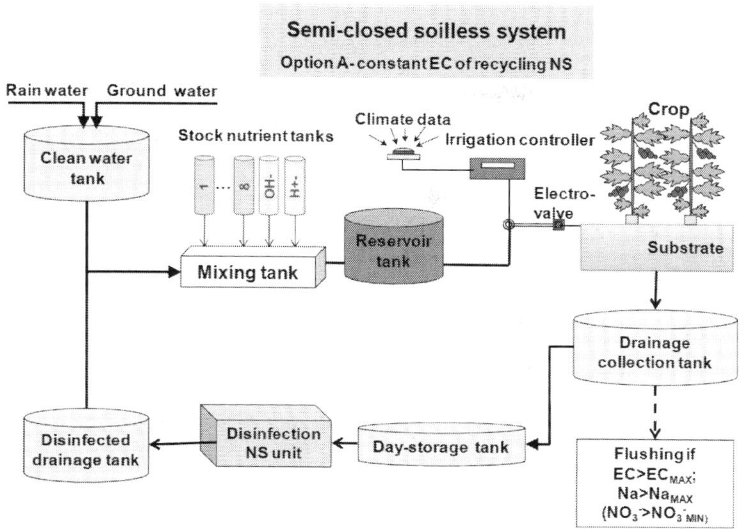

Figure 28: Fertigation scheme following Strategy A.

Source: Pardossi et al., (2011).

Therefore, recirculating NS should be routinely (every 1-2 weeks) tested by rapid tests (on farm) or by an outside laboratory to determine whether flushing is necessary and to change the composition of refill NS. When the concentration of a specific ion reaches CMAX, NS is released. According to Vermeulen et al. (2005), growers in The Netherlands are permitted to leach their systems if a crop-specific Na+ concentration ceiling is reached, such as 8 mol m3 for tomatoes. Alternately, NS is released when the concentration of some harmful substances (such as nitrate) is less than a wastewater maximum level set by law.

Strategy B:

When the EC and/or concentration of an ion exceeds ECMAX or CMAX, the recirculating NS is flushed out. WU was compensated by filling the mixing tank with nutrient solution at full strength (with the concentrations of macronutrients equal or close to the corresponding mean uptake concentrations as determined in previous studies). A rather constant level of nutritive ions is produced by this process, while the EC gradually rises as ballast ions build up. This fertigation method's main drawback is that salt-sensitive crops cannot use it since the EC oscillated in the NS.

Strategy C:

Similar to Strategy A, this process compensates for crop WU only with pH-controlled raw water for a short period of time after reaching ECMAX or CMAX in order to reduce the concentration of polluting ions in the NS prior to flushing.

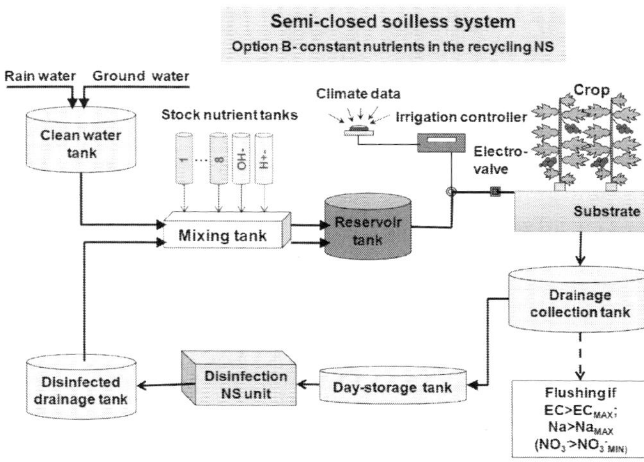

Figure 29: Fertigation scheme following Strategy B or C.

Source: Pardossi et al., (2011).

Leaching Requirement in Semi-Closed System

By taking into account fluctuations in the ionic composition of recirculating NS and supporting growers' decisions on nutrient replenishment and leaching requirements, simulation models can help to enhance fertigation control in semi-closed systems. For effective water management at the farm gate, models of W, which are determined by crop ET and LR, may be helpful tools for both growers and policymakers (for instance, for establishing limits to water and fertilizer application). There are restrictions on how much irrigation water may be used for greenhouse crops in The Netherlands (e.g., 1140 L m-2 in tomato). Several models for automatic fertigation in closed systems have been developed. However, because they need a lot of variables and parameters, practical implementations of these models are challenging. In its place, Carmassi et al. (2007) suggested a straightforward mass balance equation to forecast W of a semi-closed soilless culture based on a limited number of variables and parameters, such as CU of nutritional and ballast ions. With the following presumptions, a straightforward equation may be derived from the equation presented by Carmassi et al. (2007).

- C_U is negligible compared to the concentration of the analyzed ion in the refill NS (that is, in raw water; C_I);
- at flushing, the entire volume of recirculating NS (including the one in the substrate) is discharged.

The equation is the following:

$$LR = \frac{C_I}{C_{NS}^{MAX} - C_I}$$

After rearranging, this equation can be used to predict the highest CI that will allow a crop to be grown with known ET, tolerance to the ion under discussion, and within the restrictions of restricted water availability (represented by CNSMAX and WMAX, respectively), which in turn determines LR:

$$C_I^{MAX} = C_{NS}^{MAX} \cdot \left(\frac{LR}{1 + LR}\right)$$

Where:

$$LR = \frac{W^{MAX}}{ET} - 1$$

CHAPTER 6

EQUIPMENT IN SOILLESS PRODUCTION

CONTENTS

INTRODUCTION

A potential for higher crop output and efficiency led to the commercial launch of soilless cultivation. As part of this, technical advancements were made in relation to issues with root infections, a lack of oxygen in the root zone, fertility management, and increased irrigation plan complexity. The production of soilless container plants in outdoor nurseries became widely accepted in the 1950s and 1960s as a result of technical answers to these issues and opportunities. The development of greenhouse crops in stone wool significantly increased the commercial viability of soilless crop production in the early 1970s. The adoption of fertigation technologies, in which totally soluble fertilizers are dissolved in irrigation water to provide plants with the nutrients they need for maximum growth, is the consequence of technological advancements in irrigation and fertilization. Fertilization and irrigation have been integrated into systems that the producer aims to optimize in all contemporary production systems. Traditionally, gardeners put dry fertilizer products on the top of the root zone, causing irrigation water to carry fertilizer nutrients to the root surfaces. Systems were devised where such salts are dissolved at reasonably high concentrations in specific stock solutions once it became clear that all necessary fertilizers could be supplied through totally soluble fertilizer salts. Such concentrated solutions could be injected into the irrigation water by utilizing one or more injectors (fertigation). The technical tools used for fertigation and irrigation in greenhouse production are the main topic of this chapter. The majority of these soilless production systems fertilize by injecting soluble fertilizer into irrigation water. We also go through which crop responds best to which system and where more research is likely to lead to technological advancements in the future.

WATER AND IRRIGATION

Water Supply

Whether growing crops in a greenhouse or an outdoor nursery, soilless crop cultivation requires a sufficient amount of high-quality water. Rainwater, surface water, groundwater, and desalinated seawater are all potential sources of water. The latter three can be acquired as part of a municipality's drinking water supply directly or indirectly. Condensate from air conditioning systems is another potential source of water for irrigation in indoor production systems that use artificial lighting. The dissolved minerals and salts in irrigation water are often taken into account when assessing their quality. It is well recognized that source water with high salinity (EC> 2 mS/cm) can cause growth inhibition (and possibly other negative manifestations) for many plants. Salinity is commonly measured as electrical conductivity (EC). The accessible water supply in some parts of the world has an EC of almost 2 mS/cm (or possibly greater), which poses significant difficulties for growers. The dissolved elements in the water supply and any fertilizers that are dissolved in the water combine to form the irrigation water's EC. Sodium content of less than 0.5 mmol/L and an EC of less than 0.5 mS/cm are ideal values for the supply water. This thus makes it possible to add plenty of fertilizer ions so that the irrigation solution that the plants perceive is below the point at which issues might arise. If the EC of the source water is higher than 1 mS/cm, EC control can be particularly difficult in recirculating irrigation systems, where some soluble elements will accumulate over time. Because water quality in many parts of the world is poor and contains higher levels of sodium (Na+) and other problematic elements like SO2-4, Fe2+, or Mg2+, management of this type of situation is covered in the previous chapter, Water and Nutrient Management and Crops Response to Nutrient Solution Recycling in Soilless Growing Systems in Greenhouses. Materials dispersed in the water may not only harm plants but also the irrigation system in addition to harming them. For instance, deposits of iron and carbonates in the pipes might clog filters and cause blockages.

Rain water

Rain has always been important to agricultural production from the beginning of agriculture. Although this supply of water generally has very low EC and is hence very clean, its availability is typically irregular. Due to the uncertainties and potential production issues connected with rain events imposing ambiguity on the nutritional state of the root zone, many soilless production systems include full protection of the plants from precipitation. In order to use rainwater in soilless cultivation, it must be collected, stored in reservoirs, and then pumped under pressure (together with filters) for use in nurseries. Rainwater typically has very low EC levels in comparison to the requirements and restrictions mentioned above. Rain and wind near seas can occasionally raise the concentration of Na+ in precipitation. Rainwater that falls on the greenhouse roofs can be collected. Growers must have a storage capacity of at least 500 m3 of rainwater per ha of the production facility in the Netherlands, where there is an abundance of rain throughout the year. If the storage capacity is increased to 1500 or 4000 m3/ha, then 75 percent or

95 percent, respectively, of the necessary water can be secured in this way. Rainfall averages may range from 50 to 90 mm/month, so a storage capacity of 500 m3/ha can provide for about 60 percent of the irrigation water needs. Based on the desired capacity, this rainwater is kept in basins or tanks. For capacity greater than 1500 m3, earthen basins are employed, whereas tanks are used for lesser volumes because they are less expensive and require less production area. For a variety of reasons, certain production locations do not collect rainwater. Since the area lost to this equipment might significantly reduce the nursery's production space, it may not be economically practical to invest in collection equipment and water storage if other water sources with higher quality and lower cost can be found. Rainwater collection may not be possible in regions where inexpensive greenhouse structures are in use if the greenhouses are not built with this in mind (lacking gutters or drainage plumbing). Additionally, precipitation patterns are more erratic in many regions (such as Northern California and the Mediterranean countries), necessitating the need for gathering reservoirs that are significantly larger than those mentioned above for Northwest Europe. In some areas, aquifers can be used to store rainwater. These prevent the need for expensive space while maintaining water quality. In actuality, a lot of groundwater sources fall into this category, albeit on a much greater scale. One significant problem with using rainwater collected from roofs is that it will never be enough to meet all of the needs of most nurseries. As a result, irrigation systems must always have a backup water source. This needs to be accounted for in the investment cost.

Figure 30: Rain Water Tanks.

Source: enviro-friendly.com/water-tanks/

If tap water has been sterilized and flavor-enhanced for human consumption, it might not be safe to use in closed soilless systems. It is especially crucial that chlorides be present at less than 1.5 mmol/L if calcium and chlorine were introduced during processing. Chloride levels are often greater than this in arid and semiarid areas of the world, thus producers must control this through the use of cultivars that are resistant to salinity and other techniques. Municipal drinking water is frequently more affordable than producing such high-quality and sufficient water on-site in many parts of the world. In the absence of this, the greenhouse operation must use groundwater or surface water, sometimes in conjunction with rainwater that has been retained and irrigation runoff that has been captured.

Surface water

Water from lakes, rivers, and creeks on the surface is occasionally inexpensively and in great quantities. While this water can be of great quality if it is generally untouched by human effect, it can also be of low quality, especially if it has undergone insufficient treatment after being used for municipal purposes or because of other human environmental impacts (e.g., canal water). Salts, agricultural chemicals, improperly treated sewage, or storm water runoff from surrounding metropolitan areas could all pollute the water. Even if surface water is generally of acceptable quality, it should be remembered that water quality may not be consistent year-round, necessitating frequent testing by the grower to ensure its continuous viability for crop production. The farmer may need water treatment equipment to improve the water quality if it deteriorates at specific times of the year. Alternately, as many farmers in the Netherlands do with rainwater collected from roofs, a second supply of high-quality water can be used to mix with water of lower quality. Although effective, reverse osmosis is a frequently used treatment method that significantly raises the net cost of supply water. In an effort to create water resources for greenhouse horticulture, sea water desalination has been tested. Without government subsidies, this results in a relatively expensive price for freshwater. Desalinization technology has advanced over the past ten years, leading to the construction of large, more effective desalinization facilities in dry regions. These plants produce and supply water at a reasonably affordable cost of between $0.50 and $0.60 per cubic meter. Desalinated water consequently became a crucial source of water for domestic consumption and agriculture in places like Southern Spain, Cyprus, Saudi Arabia, the Gulf States, and other places.

Groundwater

Groundwater can be a great source of water in areas of the world with aquifers. These aquifers are often big and well-buffer such that the supply quality is quite constant. This is significant for farmers since it means that the design of the fertigation system won't need to be altered over time. It is recommended for the producer to have the source water tested every 6 to 12 months to check water quality because fluctuations in water quality can still happen. Many domestic groundwater sources haven't altered in centuries, but they're currently undergoing alterations as nutrient-rich water from agricultural operations leaches downward into the aquifer (a process that can take decades). Additionally, groundwater depletion is an issue in many regions where intensive agriculture is practiced. The intrusion of seawater into the aquifer around the coast, where agricultural output is frequently intense due to the milder weather, can lead to groundwater that is too salty to be used for plant growth (EC 10-15 mS/cm, for example). Groundwater, which primarily comes from the neighboring mountains, is frequently used in the area of Spain near Almeria (EC ranges between 0.4 and 3.5 mS/cm). The groundwater table is dropping as a result of pumping water from wells between 150 and 600 m deep, generating worries about the political and environmental fronts. A similar observation has been seen along California's Central Coast, particularly in the areas surrounding Watsonville and Salinas, where the intrusion of saline water has caused several crops to locate alternative supplies of water or shut down.

Irrigation Approaches

The machinery used to transform the raw supply water into water fit for soilless agriculture should prioritize enhancing the water's quality and applying pressure to make it flow through the plumbing. Cleaning water to make it suitable for usage typically involves the use of sand filtration. To create water pressure, pumps are also required in addition to filters. Water must be pressured in all soilless production methods in order to make uniform irrigation practical. Near the pumps, water distribution systems frequently use pipes with the biggest inner diameter. The pressure decreases with distance from the source of the water in the pipe. By reducing the pipe cross-section as the water flows through it, this issue can be somewhat alleviated. But ultimately, the irrigation water delivery system needs to be split up into circuits so that each circuit has enough pressure to ensure that the plants are watered evenly. Irrigation in Soilless Production has already covered the options of irrigating from above the plant canopy, to the top of the root zone, or from below the root zone. These irrigation strategies each require various pieces of technological gear. The image below shows a schematic overview of irrigation machinery used in soilless production. In soilless cultivation in greenhouses, irrigation and fertilization techniques are typically combined.

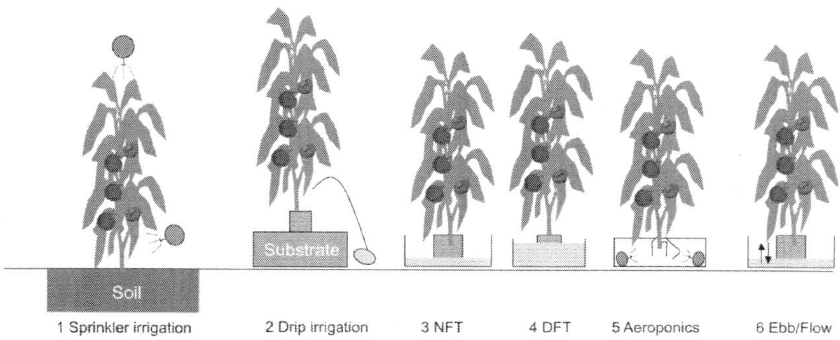

Figure 31: Types of watering methods.

Source: Van et al., (2019).

Sprinkler Irrigation

Plants in sprinkler irrigation systems receive irrigation water from above. Some shade and greenhouse buildings have piping attached above them. The piping is shielded from mechanical harm by people and cars when installed overhead, which is a benefit. Supply lines for installations may alternatively be attached to the infrastructure or buried in the ground beneath the crop. Around the world, movable sprinkler systems are also in use. Micro-sprinkler irrigation can be used to water crops whose foliage should not be wetted (such as some decorative crops). The majority of the top of the root zone receives water from spray irrigation from above, which has the advantage of requiring little initial capital and little ongoing care. Multiple sprinkler heads with close spacing between emitters can be employed to create a more uniform distribution pattern, however, sprinkler systems by their very nature have uneven water distribution. Each type of sprinkler head has a certain circular water distribution pattern. Wetting of the foliage increases the risk of plant disease development for particular crops. Micro-sprinkler irrigation systems, which provide water to the base of the plants, get around some of the limitations of overhead water supply: the crop only gets partially soaked and there is no light interference from overhead pipes. A vast number of sprinkler heads, each dispensing a lesser amount of water, are needed to irrigate all plants since the plants themselves obstruct the water's equal distribution and impede irrigation of broad surface areas.

Modern sprinkler installations are often created to meet the individual needs of the grower and the type of cropping system. Sprinkler emitters can range from rotating impact sprinklers and gear-driven ones that can throw a stream of water 2–20 meters to ones without moving parts that can produce a circular spray pattern with a radius of 1–5 meters for outdoor nursery output. Sprinkler systems are less popular in greenhouse production, and micro-sprinklers, which feed water to the base of the plants more precisely and at lower rates, are frequently used as emitters. Pressure compensation ensures that emitters only deliver water while the pressure is between the defined minimum and maximum values to prevent the emitters from leaking. Emitters have also been developed with various characteristics that enhance performance. Sprinkler systems are the most commonly employed system in the manufacturing of outdoor containers due to their low cost when compared to other systems. However, farmers normally only employ sprinkler irrigation with closely spaced plants in these kinds of businesses. In order to minimize the quantity of watering that completely misses the plants when they are placed far apart, drip or micro-sprinkler irrigation systems are often used. Due to the aforementioned disadvantages, sprinkler irrigation is less frequently utilized in greenhouse production.

Drip Irrigation

In soilless culture in greenhouses, drip irrigation is currently the most popular irrigation method. There are two types of drip systems: in-line systems and microtube systems. The former usually entails a large number of emitters and supply tubing to enable an emitter to distribute an irrigation solution at a precise spot, typically close to a plant's base. The latter consists of supply tubing that has emitters affixed to it directly or encased within it. The system may function with relatively high pressure (at the pump or source), decreased to uniformly lower pressure at the emitters, thanks to drip irrigation with microtubing (capillaries) and drip emitters. The grower can maximize the uniformity of irrigation solution delivery to each plant using this method. This is crucial for crops where each plant has a low planting density or a high value, making it possible to affordably maximize fertigation of each plant. In order to ensure that each emitter releases the irrigation solution at the proper location, they are often attached to either a small stake or a weight. According to the needs of the crop and substrate, emitters or nozzles with different flow rates and working pressures are available and can be selected. Accordingly, emitters with low flow rates are often utilized with substrates that have high infiltration and drainage rates. This is because emitters are typically matched to specific types of substrates. To ensure a uniform distribution of water, pressure compensated emitters can be employed. Leakage-protected emitters ensure that water is only supplied during irrigation events and prevent dripping at various points throughout the irrigation system (Fig. below). By preventing the irrigation system from draining in between irrigation events, the system also starts up more quickly during an irrigation event, improving irrigation uniformity, especially for short-duration irrigation events.

Figure 32: Leakage-protected emitters to water plants in soilless cultivation.

Source: Van et al., (2019).

In-line drip irrigation can be used to water crops in beds where plants are growing at a specific spacing by balancing the planting density with the fixed in-line emitter spacing. The same characteristics as those of ordinary emitters may apply to in-line emitters.

Nutrient film technique

The Nutrient Film Technique (NFT) involves growing plants without a substrate by keeping a layer of nutrient solution on the roots. When NFT originally came out, it appeared to be the perfect growing system because it provided the best control over root watering without the need for a substrate. Since it was challenging to create optimization algorithms, NFT is now only employed for a few particular crops. Technically, most crops might be grown in an NFT system, but this has not been widely adopted, likely because these systems are not able to withstand even a little stoppage in the supply of water and nutrients and there is a high danger of the development of root-borne diseases. The plant's roots are located inside the bottom of a trough that has a slope of 0.3 percent to 2 percent. At the elevated end, a nutrient solution is continuously delivered so that it flows through the trough at precisely the pace necessary to maintain the roots thoroughly moist. The solution is permitted to drain at the bottom of the trough. The layer of nutritional solution should be virtually film-thin. The diameter of the trough varies depending on the crop; troughs of 4–8 cm are adequate for plants like chrysanthemums (*Dendranthema grandiflora*) and lettuce (*Lactuca sativa* spp.), but a trench of 15 cm is required for tomatoes (*Lycopersicum esculentum*) and sweet peppers (*Capsicum annuum*). The trough's length ranges from 1 to 20 meters. Various types of materials, including polyethylene liners, polyvinyl chloride (PVC), polypropylene, and coated metal, have been employed depending on the crop and the size of the troughs. Crops like chrysanthemums and lettuce have been found to benefit from water flow rates between 3 and 8 L/m2/h. Vegetables that bear fruit gain from a faster flow rate. If the flow rate is too low, the issue is not a lack of water but rather a lack of nutrients, particularly for plants whose roots are downstream in the trough and are exposed to water from which many other plants have already extracted some nutrients. In an NFT system, the slowest flow rate that is sufficient to keep the roots coated with water may not be sufficient. The plants near the back of the line receive the least nutrients, particularly potassium. There may be a difference in the flow rates required for a young crop (2 L/m2/h) and a mature crop (5 L/m2/h). The plants that will be used in NFT systems are grown in tiny pots and then transplanted into the trough once they have developed a strong root system. A tissue lining in the bottom of the trough can be utilized to reduce this issue in circumstances when the water flow meanders in the trough, skipping some plants. Shaped troughs, such as V-shaped ones, also avoid this issue.

Deep flow technique

A technique known as deep flow technique (DFT) makes an additional effort to keep the roots constantly exposed to flowing water and nutrients. While in NFT the water film is as thin as possible, in DFT the nutritional solution is continually flowing at a depth of roughly 5 to 15 cm. Controlling the nutrient solution is made significantly easier by the substantial water and nutrient buffer. The amount of water and nutrients that are actually absorbed by the plants is a very small percentage. The technique is useful in areas where nutrient solutions temperature swings might be a concern because of the vast amount of water that buffers the temperature. A DFT system's troughs are typically between 100 and 130 cm wide. Polyurethane foam is used to hold plants in place in holes on polystyrene panels, which float on the water or rest against the sides of the troughs. In order to make it simple to plant and harvest crops like lettuce or herbs, the system is frequently built at working height.

Aeroponics and aero-hydroponics

In aeroponic growing systems, plant roots are suspended in a space where a nutritional solution is sprayed over them continuously or sporadically by emitters. The design is comparable to DFT (closed square box, roughly 1.2 m wide and 5–10 m long), but there isn't a water layer; instead, the roots are continuously misted). A triangle made of polystyrene boards creates a similar enclosed environment as a Japanese commercial design and as a research arrangement. Around the roots of the triangular structure, oxygen availability is higher and the relative humidity is 100%. Care should be taken to ensure that the higher plants get adequate water, though. On occasion, a thin film of water forms on the bottom, acting as a barrier for the plants. Kratky (2005) used increased lettuce yields in contrast to a continuous water layer surrounding the roots to demonstrate the value of the additional oxygen present in these systems. In the literature, the amount of water provided to the plants is not specified; a timer is frequently utilized, but the nozzle's release rate is not. The lack of a root-buffer for water, nutrients, and heat is a drawback of both NFT and aeroponic systems. By merging aeroponics and DFT, Soffer and Levinger (1980) created a hybrid system they named "The Ein Gedi System" and popularized the term "aero-hydroponics." In this system, instead of just a thin coating of the nutritional solution, the roots of the plants extend through a panel and dangle in the air like in aeroponics. This technology differs from other soilless production systems in that the bulk flow of water and nutrients bathes the roots faster and the nutrient solution is continuously aerated. All forms of aeroponics rely on power to transport the water in large quantities, which is essential to the plant's existence. The improvement in food uptake and oxygen availability must be balanced with this energy cost in order to assess the commercial viability of such a system.

Figure 33: Aeroponics.

Source: nextpittsburgh.com/environment/

Ebb and flow/Ebb and flood

Container-grown plants are flood irrigated on a watertight tray or floor in "ebb and flood" (EF) systems. During irrigation, water flows onto the tray or floor until it is flooded and the base of each pot is submerged. The substrate inside the containers' hydraulic conductivity determines how long the flooding must last. If the substrate has a high unsaturated hydraulic conductivity, the time should be sufficient to allow the water in the root zone to be wicked to the top of the root zone, which is typically 10 to 30 minutes (longer if not). The substrate can then drain after the tray or floor has been emptied. These systems are designed with sloping surfaces and drainage channels or grooves to maximize water homogeneity over a floor or tray amongst all pots/containers while promoting quick drainage. To avoid root malaeration and infections, it is especially crucial to enable the water to drain entirely. Algae growth, root infections, and uneven watering will happen if moist spots persist around the plant.

FERTIGATION EQUIPMENT AND TOOLS

Water performs two crucial roles in the growth of plants: it supplies a necessary resource for the growth and also transports nutrients. Fertigation is a term used to describe irrigation techniques where both of these functions are actively combined in a single system using entirely soluble nutrients. The dynamic regulation of the delivery of nutrients in fertigation systems is the main topic of this section. A diluter/dispenser unit mixes irrigation water with fertilizers as concentrated chemical solutes from two to eight stock tanks. The resulting nutritional solution is delivered to each individual plant via a series of pumps, valves, and irrigation capillaries. The amount of irrigation is frequently managed by the supply's time-duration operation or by a succession of a predetermined number of controlled volumetric additions. The fertilizer diluter/dispenser unit is connected to the greenhouse's numerous irrigation sections through valves. There are about 50 section valves available, and each one has the equipment to measure and record the amount of supply and drainage as well as daily averages of EC, pH, and water use. Supporting initiatives take care of the storage for rainwater stock, drain storage, and storage for cleaned drain water. Selection criteria identify the source of the clean water when more water is required in the system. Water from various sources shouldn't be combined beforehand because pure water is a very scarce resource.

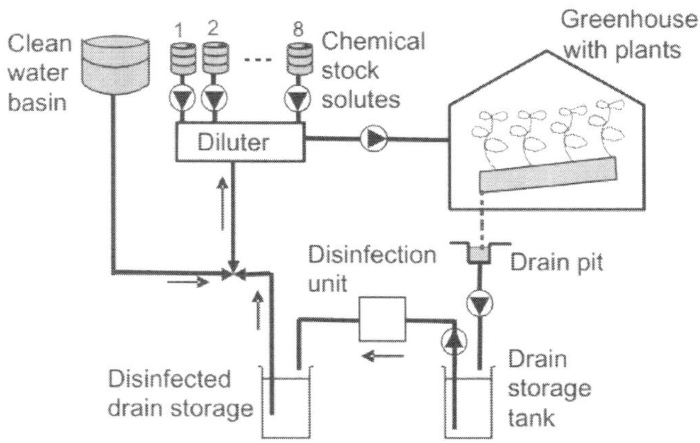

Figure 34: Nutrient solutions pathway in a closed production system.

Source: Van et al., (2019).

The drain water in enclosed growth systems is recycled. When extremely concentrated stock fertilizers are combined before dilution, crystalline salts may precipitate. To obtain a predetermined EC value of the input water to be given to the plants, recycled drain water will be combined with freshwater in a chosen proportion. The controller of the device should include instructions for automatically cleaning the filters in the recirculation circuit as well as an exact control of the EC and pH of this input or supply water. These methods should be part of a thorough and thorough fertigation management plan to prevent failures brought on by unforeseen EC and pH values or system blockages brought on by precipitation. It is possible to install EC and pH sensors inside the control devices. Both types of sensors can be employed in horticultural practice for dynamic control and compared with the findings of laboratory investigations to confirm the calibration. The content of the nutritional solution is also determined using relevant laboratory analysis. It is possible to identify a specific ion dissolved in a solution using ion-selective electrode (ISE) sensors. The activity of cations and anions in the root environment is measured using ISEs. The majority of macronutrients, including K+, Ca2+, NO-3, SO2-4, NH+4, as well as Na1 and Cl-, can be detected with ISE sensors. The measurement that results from using manual test equipment gives the grower strong input on how the nutrient solution is changing. However, installation into dosing systems is not currently practical due to this technology's regular (daily) calibration requirements.

Stock Solutions

According to the chemical makeup of the source water, a special recipe of nutrients can be created for the particular crop. A system was devised by Verwer (1978) with a single central tank to contain irrigation solution made by combining water with two stock solutions, referred to as A and B solutions, in 1970, when hydroponic production was introduced in The Netherlands. Even now, several systems of this kind are in use. The two stock solutions are created at concentrations that are either 100 or 200 times greater than the typical concentration of the solution that will be given to the plants. All calcium compounds are located in tank A. All sulfates and phosphates have been dissolved in container B. The fertilizers in the two stock solutions would precipitate in the pipes, valves, and emitters if they were combined in these high quantities, which would be devastating. The solutes from the A and B tanks are diluted with freshwater in the mixing tank to obtain the proper concentration and EC level of the nutrient solution, which helps to prevent precipitates. The pH of the batch of nutrient solution in the mixing tank may be controlled by adding additional amounts of an acid or basic solute. The A and B blends' recipes are chosen in accordance with the needs of the crop that will be fertigated and are based on twice-monthly laboratory analyses. Normally, fertilizer is manually applied to the tanks A and B. Another strategy is to distribute specific liquid fertilizers that have been purchased from a supplier and kept in specific stock tanks. Depending on the supplier, a standard mix for the majority of commercial greenhouse crops combines 6–10 liquid fertilizer concentrates. Each concentrated fertilizer solution has a separate tank on the delivery truck. A consistent discharge technique guarantees that every chemical fluid is placed in the appropriate container, reducing the possibility of mishaps. There is no longer a need for further manual labor. The expense of using this technology, which involves moving a lot of water, is a serious drawback.

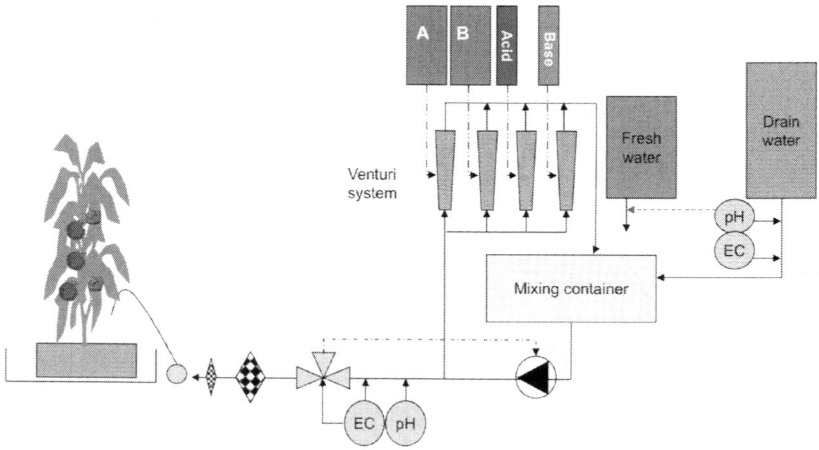

Figure 35: Fertilizer diluter/dispenser unit.

Source: Van et al., (2019).

Only places with a large number of greenhouse operations, such as the Netherlands, the area around Almeria, Spain, or the Brittany region of France, may use this technology profitably. Another method is to manually combine precisely weighed soluble fertilizers from bags of typical fixed combinations in the A and B stock tanks. By using 1 or 2 m3 large plastic bags, where the provider has pre-mixed the solid fertilizers into A and B bags in accordance with the most recent laboratory analysis, the amount of labor can be reduced. The bags are supplied on regular pallets, are readily emptied through an opening at the bottom of the bag, and are then set on a platform above the A and B tanks by a forklift truck. To make the stock solutions, the entire contents of the huge bags are utilized at once and combined with precisely the right amount of water. In order to ensure that changes in the composition of the A and B solutions occur simultaneously, fresh recipes for the A and B stock solutions are prepared sequentially but released into both stock tanks at the same time. The nutrient solution's recipe must be modified to meet the crop's requirements as it progresses from the seedling stage to the reproductive stage. A regular mix specific to the species is used to begin cultivation.

A test is then conducted every two weeks on a sample collected from the root environment, and this recipe is changed in accordance with recommendations from a professional laboratory. The recommendations take into account the makeup of the various stock nutrients, whether they are solid or liquid stock fertilizers, and details regarding the crop's condition. On the basis of this data, the necessary recipe modifications are determined. The diluter/dispenser unit cannot provide all potential combinations of these calculations as recipes. Fertilizers have evolved in recent years from straightforward single- or multi-element fertilizers to increasingly complex combinations of elements, tailored for a particular growth strategy. By adding acidic or basic fertilizer solutions, the option of altering pH levels in liquid fertilizers has also been offered. Computerized calculations have replaced manual calculations in the formula, allowing for the consideration of a wide range of restrictions. These days, the planter can even directly access these algorithms using a spreadsheet or a specialized computation tool linked to a specific fertilizer brand. Certain mixtures might result in precipitation or ion combinations that can't be made from the stock solids or solutes of the available fertilizer. The dilutor/dispenser units may experience precipitation, blockages in the transport conduits, or even explosions if they are forced to follow a "impossible recipe." Additionally, optimization methods may occasionally result in a more economical selection of comparable fertilizers.

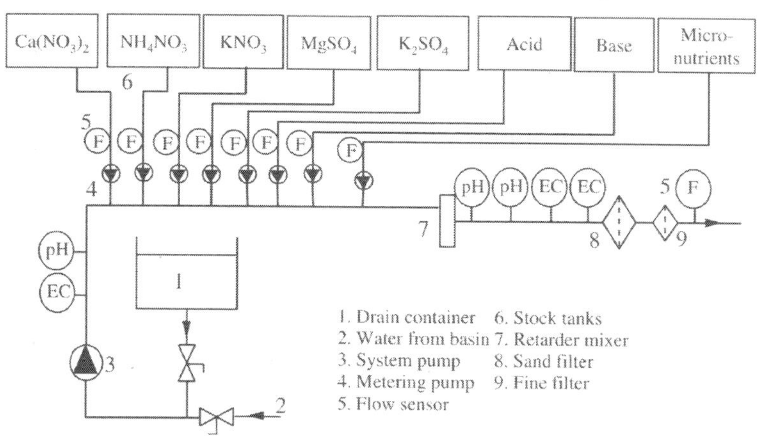

Figure 36: System with metering pumps.

Source: Van et al., (2019).

Commercial Injector Systems

***The mixing tank system*:** The following are the operational principles of mixing tank diluters: Connected to the several fertilizer stock tanks is a mixing tank. Nutrient injector systems are used to feed quantities of the highly concentrated fertilizer solutions into the mixing tank, where the nutrients are diluted with fresh water to a precise concentration per recipe before being pumped to the greenhouse plants. The highly concentrated stock mixtures can be added in a number of different ways to the mixing tank:

- Metering pumps deliver the stock liquids into a main supply line or an open, unpressurized mixing tank. In the mixing tank, one batch is prepared at a time, and then it is pumped to the greenhouse. Early designs of computerized nutrient fertilizer systems contain examples of these systems. Commercial farmers currently have a small mixing tank of around 2 m3/ha and a day storage tank of 1050 m3/ha. Due to its modest size and ability to generate fresh solutions quickly, the diluter unit can fill the storage tank from which irrigation water is drawn.
- A pressure decrease caused by a Venturi fitting inside the main supply line drives the stock fertilizers into the mixing tank or premixing chambers.
- One tube for each concentrated stock solution is positioned slightly above a mixing tank in small container tubes (B5 L). A precise amount of stock solution is released into the mixing tank by a valve at the bottom of the tube. The volume of stock solution released is measured by a pressure sensor at the tube's base. Simple pumps or gravity are used to fill the container tubes when the stock tanks are hoisted.

Despite the existence of all three ways, the first one, which uses 100* concentrated fertilizers in containers A and B with an additional day storage tank, is the one that is most commonly used in commercial practice. By adjusting the degree of nutrient and water dilution, the EC of the supply solution can be changed. Simple systems, such those that combine four nutrient stock fertilizers with an additional acid or basic solution for pH correction, are designed to mix a small number of stock solutions. To ensure safe functioning, double EC- and pH-sensors are added. The mixing procedure, the alerts, and operator communication are all managed by a computer system. Systems with overhead sprinklers, drip irrigation systems, or ebb-flow systems all employ mixing tank dispensers. They can be set up for both batched by-pass mode and in-line mode. The systems come in a range of output capacities, from less than 1 m3/h to more than 40 m3/h. There are only a few (four to six) conceivable variations in nutritional formulations. The EC set value can be adjusted to reflect variations in the radiation experienced worldwide, however substrate systems seldom ever benefit from this. It takes too long to switch from the older, higher-EC solution to the new, lower-EC solution (if there is more radiation). The majority of EC steering is done in favor of vegetative or more generative growth, with less frequent irrigation turns being used to offset daily radiation influences.

***Direct injection system*:** An innovative and appealing replacement for the use of metering pumps is the injection of concentrated fertilizers by way of a pressure decrease in the main supply line. A Venturi is used to create the pressure drop. The concentrated fertilizer fluids are sucked in due to this pressure reduction. It is a more efficient way to transfer fertilizer since it has fewer gadgets and moving components that need upkeep than a system with metering pumps. Units for direct injection dilution and dispensing come in a range of sizes and capacities. Simple direct injection systems use separate premixing chambers in which highly concentrated stock fertilizer solutions are mixed with supply water. This allows the stock fertilizers from the A and B tanks to be mixed and diluted in the mixing tank. After being diluted to the proper concentration for the plants, this solution is subsequently injected into the main supply line. More complicated systems are used to mix and dilute highly concentrated stock solutions of different liquid fertilizers. For each stock tank, the dispenser unit employs a fertilizer channel. Once more, the main supply line's Venturi pressure decrease is what causes the concentrated fertilizer fluids to travel. A tiny ceramic inlay with an opening is used to seal the channel. The orifice guarantees a constant and constrained flow rate. The various channels are filled with ceramic inserts that have various orifice diameters. The cumulative flow over time is carefully controlled by a flow meter and a fast-switching valve. The mixing chamber is connected to four channels. Pulse activating the valves controls how much fertilizer is carried into the mixing chamber. With the help of a static retarder in the main supply line, dangerous combinations of highly concentrated stock fertilizers and leftover nutrient ions in the supply water are avoided (recirculated in case of closed systems). The measured EC value of the supply water is fed back into a computer controller to regulate the mixing and dilution of the different specific fertilizers.

The controller determines and adjusts the amount of each individual fertilizer based on the detected water supply flow rate and the volumetric relationship between the numerous fertilizers, as specified in the recipe chosen by the grower. When the mixing algorithm is appropriately used, the solution will have the desired pH and EC. The necessity for fine-tuning the pH and EC value is introduced by other factors, such as the usage of recirculated drainage water or slight variations in the clean input water of the system's composition. By automatically modifying the output of the relevant individual fertilizer channel in accordance with its recipe value, this is accomplished. The next cycle of fertilizer supply can remedy an overabundance of a specific nutrient ion. The concentration of each stock solution is taken into account while calculating the fertilizer recipes as an input parameter. Thus, the system can be simply modified to accept any concentrations of stock solutions that are now accessible, such as common industrial solutions or unique stock solutions made specifically for horticulture. Inherent reliability is made possible by strong materials and a limited number of moving parts. In general, eight liquid fertilizer fluids—including one channel for micronutrients—are sufficient to implement every conceivable recipe. There are injection dispensers with capacities ranging from 100 m3/h to more.

Figure 37: Direct injection system for single-element liquid fertilizer.

Source: Van et al., (2019).

Safety and Protection

It can be dangerous to combine liquid ionic nutrients that are extremely concentrated. The system must include additional safeguards to shield the farmer and the personnel from potentially dangerous circumstances. For this reason, additional sensors should monitor the diluting, mixing, and dispensing process for potential dangers including;

- Excessive or insufficient flow in the individual fertilizer channels,
- Liquid fertilizer leakages,
- Intake of air into the fertilizer mixing channels,
- Unexpected drop in capacity, and
- Jamming of the channel.
- General checks on the following should also be made:
- Sudden or large deviations from the expected EC and pH values,
- Leakage from the stock tanks (also when the equipment is switched off), and
- The stock tanks should be stored in a containment room with the capacity to house all the desired stock tanks.

PRODUCTION SYSTEMS

Different sorts of production systems have been devised to grow particular crops since the usage of soilless culture was first introduced. Typically, the crops produced in this manner are "high-value" crops with high rates of economic return, assuring that the planter will profit from the input-intensive system that has been implemented. In the early stages of this sector, manufacturing took place underground in heavily altered soils; above-ground soilless systems are a more recent innovation. There is no relationship between plants and soil in today's soilless production methods, some of which are lifted above the ground and others of which sit directly on the ground. We will cover the benefits and drawbacks of each of these production methods as well as their applicability for particular crop categories (crops in rows, crops in beds). Although there won't be any discussion of substrates in this section, there will be a distinction between solid substrates (like stone wool or polyurethane slabs), granulated substrates (like perlite, pumice, coir, or peat), and none at all (as in NFT and aeroponics). The majority of systems can be classified as either open [all the leachate flows into the subsurface without recirculation or is used in other (field) crops] or closed (the leachate is reused).

Systems on the Ground

Ground preparation

When constructing a production system that will be placed directly on the ground, it is crucial to prepare the ground beforehand to reduce the likelihood of weeds and diseases. To guarantee a disease-free start and stop the germination of weeds, you can employ steam sterilization, a chemical treatment, or a solarization treatment. In most cases, treating the top 10 cm of soil is adequate. In order to properly manage the drainage water from production systems that are directly on the ground, the ground must also be prepared to satisfy the system's hydraulic requirements. If the root zone is not completely horizontal or inclined evenly, wet and dry zones will form. As a result, the plants' access to water won't be consistent, which will cause oxygen deprivation in some regions and root rot in others. EC levels may vary as a result of deviations from this, which would impair output and quality. Most leveling is done along the span of the greenhouse at a slope of between 0.3 and 0.5 percent. To prevent significant height changes within the greenhouse, the whole length is divided into portions that are 20–25 m long. After leveling, the subsoil can be suitably shaped using a molding machine. Small canals are dug to accommodate the drain water collecting tubes, and the substrate bed that will serve as the root zone may be lower or higher than the aisle. For larger regions, hand shaping is impractical and the accuracy of the job is likely to be subpar. If the manufacturing system is heavy, soil compaction may also be necessary. In order to control weeds and too much moisture, a plastic barrier is typically placed between the production system and the ground. This can be a plastic film or a sturdy woven fabric with a fine mesh (anti-root foil). The advantage of using a light-colored liner is that it reflects light back up toward the plants. This is crucial in latitudes with lengthy dark periods and brief sunshine hours. A material that is light in color (such as white) has the drawback of generally allowing enough light to pass through it and into it, which can lead to some weed germination and UV deterioration of the plastic. In places where it is more important that the earth beneath the liner reach somewhat higher temperatures (cold greenhouses without access to nighttime heating), a black liner may be more useful than a white one since it will retain more heat during the day and release it at night. Additionally, a dark substance typically withstands UV radiation's deteriorating effects better. A liner also stops plant diseases from spreading from the ground into the manufacturing system. When water needs to infiltrate into the ground beneath the aisle and production system, woven material should be utilized rather than impervious film. In these situations, it is important to prepare the ground so as to prevent overly compacting the area where the aisle will be. When wet, the woven material generally becomes less slippery than the impermeable sheet. The design of plant layouts inside ground-based systems varies significantly. Some production techniques use long rows of separate plants spaced at a specific distance between them, or "single rows." Other approaches create a "bed" of plants by placing multiple plants in a row next to one another. The term "single row" is used in this chapter to refer to a line of distinct plants placed in a substrate or trough to generate a root zone that is 4–25 cm wide. On the other hand, a "bed" is described as a group of plants that are arranged in a bigger space (>1 m), with the root zone completely covered by substrate.

Crops in single rows

Crops including roses, cucumber, tomato, pepper, strawberry, and lettuce are grown using single row methods. Small quantities of substrates—usually in the form of pressed peat pots, blocks, or plugs—are typically employed during propagation for crops grown in close-cropped rows. These are used to start propagules, either from seed or cuttings, which are then moved to a production system. When a single row is wide enough to allow for harvesting and plant maintenance to be done from one side of the row, single row systems are frequently set up so that two single rows are next to one another (creating pairs of rows). As a result, a significant portion of the aisle area is efficiently eliminated, increasing total production efficiency. In order to keep irrigation water contained, the soilless substrate that makes up the root zone of plants cultivated in single rows is typically wrapped in plastic film. Preformed materials typically come in bags with enough support to allow them to be placed directly on top of the liner or in the troughs without the requirement for a reinforcing container. Examples of these materials include stone wool slabs, coconut fibers, and some granular substrates (perlite). The root zone needs to be physically contained in substrate products that are not in these forms in order to maintain the correct shape. This can be done using troughs, buckets, sacks, or nursery containers. Directly on-the-ground production systems are typically more challenging to design as closed systems. In an open system, drain water seeps into the soil beneath the crop straight from the root zone through the liner. This conventional method has the advantage of requiring little investment in resources, but it wastes a lot of water and fertilizer and pollutes the environment. The permissible amount of nitrate leaching is now governed by government laws in the majority of Europe and North America, and open systems are being phased out in many of these regions. This has sped up the transition to closed systems. The troughs that collect the drainage water from slabs and bags have been made from a range of materials, including coated metal troughs, U-shaped polypropylene troughs, V-profiles, and PVC profiles. One of the least expensive closed systems has collection pockets on one side and is covered in a plastic bag or film. To maintain the drainage channel open, a drainpipe is frequently installed into the channel along the length of the row. A system of drainage troughs can be built using polypropylene or polyethylene sheeting that is 0.5–1.5 mm thick and delivered in rolls. These are usually 15 to 25 cm wide and have 5 cm high vertical edges. The base might have a flat or V-shaped shape. For the system to work well, the ground must be leveled and compacted to avoid depressions where water can pool.

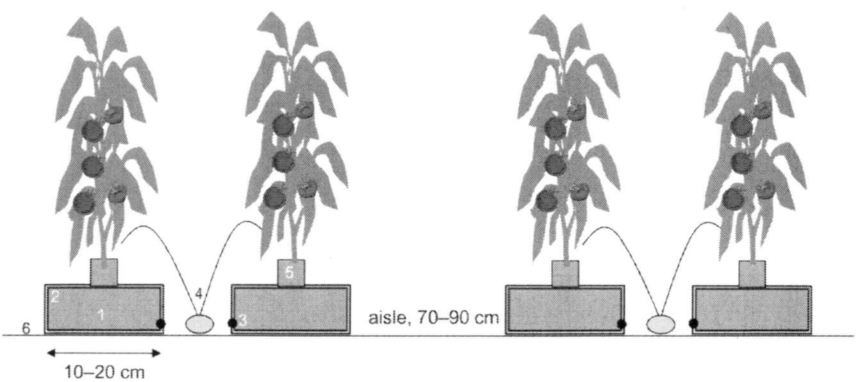

Figure 38: Open systems for single row crops.

Source: Van et al., (2019).

- Substrate slab or granulated substrate
- Foil around the substrate
- Drain holes are made 1–2 cm above ground level and at the non–aisle side
- Drip irrigation at the non–aisle side
- Pots for plant raising, using the same substrate
- Plastic liner covering the soil

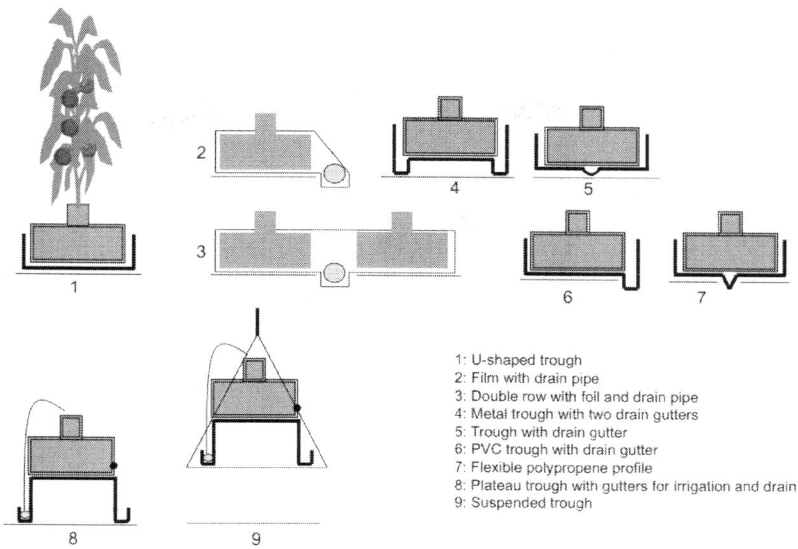

Figure 39: Troughs and profiles for closed single row systems.

Source: Van et al., (2019).

Because specific connectors connect the troughs and the main pipeline, recirculating the fertilizer solution to the water treatment unit is simple. PVC has also been used to make troughs, but these designs are far less adaptable. This material cannot be transported to the greenhouse in roll form in order to produce the trough there and then because it is brought in its final shape from the factory. The appropriate configuration must be put together at the greenhouse from lengths of 6 to 12 meters. Breakage is a possibility. The material does have the benefit that the desired connections can be simply sawed and bonded. Additionally, metal troughs can be produced using either on-site shaping from huge rolls (using a machine) or from pre-shaped trough pieces (max 12 m, depending on truck size). It is crucial to remember that some subsoils (peat or clay soils) may not have adequate support for a plastic system, so stronger metal systems are needed to prevent breakage and reduce drooping. Cost may be a deciding factor in whether plastic or metal needs to be utilized. The requirement to clean and/or disinfect the troughs between crops is one factor to take into account when choosing a material for drainage troughs. A trough made of thin plastic is liable to deform if steam or hot water is poured into it (under high pressure). These materials can be solarized, but in order to kill viruses, temperatures must rise to at least 70°C, which obviously depends on the local climate. Although PVC is rarely used for single row crops, its strong structure makes cleaning it relatively simple. It also has the drawback of being illegal in many nations, which creates a disposal issue. Burning during disposal releases chlorine into the atmosphere. After use in greenhouses, polyethylene and polypropylene can be totally burned without producing too much pollution. With high-pressure equipment or chemicals, it is very simple to clean or disinfect hard material in greenhouses. When the soil structure of the ground causes a production system installed directly on the ground to sag, a system made up of both a coated metal trough and plastic film may be required. Coated metal troughs come in a variety of widths and shapes, can be positioned on or above the ground, and can even be moved around. Although a metal trough system is typically more expensive than a plastic (polypropylene) trough, it may be more durable and last longer—up to 10 years—and is less prone to the ground settling and sinking. The harvest labor conditions are greatly improved by suspending the troughs, but this method necessitates a robust greenhouse design. In addition, the metal troughs can be supported without regard to the design of the greenhouse. It is not advisable to lay the substrate directly on the ground without first collecting the drain water for financial reasons. The (expensive) fertilized water can be used again if the drain water is gravity-collected via troughs into a drain tank (another greenhouse or open field crop).

Vertical systems

Some plants in vertical systems grow on the ground, while others are piled vertically in rows. These methods have received mixed results when used for crops with small plants, such as lettuce and strawberries (Fig. below). The substrate bag is filled with the nutrient solution, which is applied at the top and drips through it. The bottom has a collection of drain water. Aeroponics, stacked containers, and containers fastened to a pillar are all examples of similar systems in operation. The major motivation behind such a strategy is to utilize the available floor space as effectively as possible in an effort to increase the yield per m2. In a greenhouse, however, the intensity of the light is divided over a greater number of plants, giving the plants at the bottom less light than those at the top. The effects of this on year-round cropping should not be overlooked: if replanting is started only after the bottom plant is ready, each layer will yield a high-quality crop, but it will take longer before all plants are harvested and the system can be replanted. In contrast, if harvesting of all plants starts as soon as the upper ones are ready, the top layer crop is of good quality, but the yield from the lower layers decreases. A-frames can be stacked similarly to increase yield per square meter of greenhouse space. Despite this issue, vertical methods for producing Belgian endive are practical. Plants grow from seed to a mature root that may be lifted and stored cool in the open field for six months (21°C). A specified number of roots are placed in containers outside the growing area every week, as closely together as feasible. These are subsequently put in 4–8 stacks of 100*120-centimeter containers. From top to bottom, the nutrient solution can flow through the containers in a zigzag pattern before being recycled. The plants receive water, are maintained at a temperature of about 22 °C, and are shielded from light. After a period of 3–4 weeks, Belgian endive can be harvested for its white heads. For asparagus, a comparable strategy has been used.

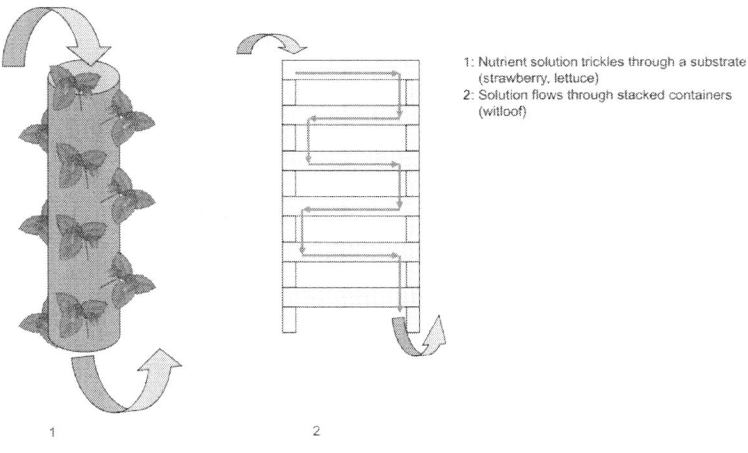

Figure 40: Vertical systems.

Source: Van et al., (2019).

Plant factories with artificial light, or PFALs, are the product of recent advancements in vertical farming and are enabling the development of urban farming systems that only employ soilless culture. Here, a variety of lush green plants like lettuce or herbs are grown in hydroponic systems in numerous layers atop one another using just artificial illumination and no sunshine at all. In most cases, the lamps are not the same as those used in greenhouses; instead, they are LED lamps, which emit less heat to the plant and may therefore be positioned near the foliage. Since every lamp produces heat, the air must be continuously cooled by air conditioning. As a result, this production method is more expensive because it uses more energy. The return is also higher because all variables can be optimized in ways that aren't achievable with other production systems (more biomass per unit area and time). As a result, it is crucial for the producer to perform an economic analysis to determine whether the extremely intense farming operation would provide a profit. Prior to the start of this process, a market must be established, and this market must have a higher market price than the typical (bulk) produce. The levels of investments are substantially higher (equipment, artificial lighting, and manpower), and these levels are typically not offset by transportation cost savings. Careful management of the air in which the plants are growing is another essential component of PFALs. There is no need to use pesticides because this air has been thoroughly filtered to remove all pests. In addition, plant development can occur around-the-clock as long as the environment contains the highest CO_2 concentration possible. A crucial regulated variable in PFALs is humidity, which has a direct bearing on how the hydroponic system functions. In order to maintain high concentrations of dissolved oxygen, air bubbles in the nutrient solution also produce humidity, which needs to be eliminated. As a result, the air conditioning system needs to be built to handle both the temperature and humidity loads at once. Condensate water's ability to be recycled is a crucial aspect.

Crops in beds

Chrysanthemums, carnations (*Dianthus caryophyllus*), freesia (*Freesia* spp.), and alstroemeria (*Alstroemeria aurantiaca* L.) are a few flower crops that are cultivated in beds with a density of 20–60 plants per square meter. Since the return on investment is insufficient to support such a system for these crops, row systems with slabs or substrates in bags are not used in this situation.

***Polypropylene liner and substrate*:** A 110–140 cm wide bed with a 1-1.5 mm thick polypropylene liner between the subsoil and a layer of 5–10 cm of a granulated substrate makes up the most common system for beds-grown crops. When the system is above the subsoil, sidewalls are required. The aisles can act as sidewalls if they are raised. At the base of a bed, a drainage pipe is placed. Heating or cooling pipes might be buried, depending on the type of crop. This method is particularly helpful for crops like lilies (*Lilium longiflorum* Thunb.), freesia, and alstroemeria, which must be completely removed between crop cycles. There is yet to be a breakthrough for these kinds of crops. Since Europe has a small geographic region and a wide range of systems, it appears that people are still hunting for the optimal system.

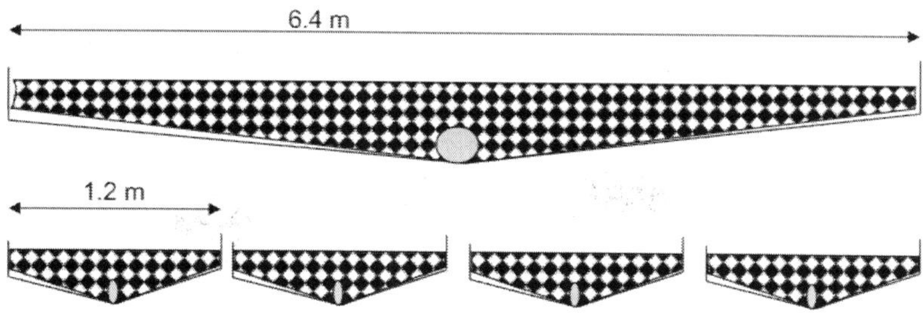

Figure 41: Systems for crops in beds.

Source: Van et al., (2019).

***Polystyrene panels*:** NFT, DFT, or aeroponic systems can be used to grow crops in liquid culture hydroponics with the plants installed inside a polystyrene panel. Young plants are supported on polystyrene panels that have been specially designed for use in plant production and are rooted in a synthetic substrate (mineral wool, urethane foam, etc.). In an aeroponic system, structural panels make up the top of a chamber where misters continuously sprinkle the roots with nutrient solution at a frequency that maintains them moist. The polystyrene sheet in this application needs to have a strong enough structure to support the fully developed plants. As a result, huge plants like greenhouse tomatoes are typically not grown using this method. In DFT systems, the polystyrene panels can float on top of the water with plants planted as above, allowing the roots of the plants to reach the nutrient solution. In order to ensure that the leaves of fully developed plants do not come in contact with the nutrient solution, a polystyrene material of sufficient thickness is utilized. In many DFT systems, mounting the plants at a constant height above the water level is an alternate strategy. In such installations, the level of the nutritional solution must be carefully managed to ensure that the roots are always moist. In addition to greenhouses, plant industries also employ this application. In order to save labor expenses, fully mechanized liquid culture hydroponic systems enable a high degree of automation for such systems. DFT systems have sizable ponds built inside greenhouses that are.15 cm deep in nutrient solution, and the plants travel through the greenhouse on polystyrene panels floating on the nutrient solution. On one side of the production area, floats containing immature plants are lowered into the nutrient solution, while existing floats are pushed toward the other side, where panels with harvestable plants are removed by workers who pack them for storage or shipping to markets. The pond is typically entirely covered with panels, ensuring that the number of panels taken out during harvesting equals the number of panels added at the other end. Algae growth is reduced by the panels' total cover. The harvest rate determines how quickly the panels move. There are other less expensive methods, like a box with a lid. The cover is around 50 by 300 cm in size, and the nutritional solution's composition is not regulated; instead, it is simply topped off if the water level falls too low. An innovation recently was a DFT system for leek cultivation (*Allium porrum*). Here, the aim was to grow a white shaft out of the ground that was at least 14 cm long (to meet market demand). Finally, a floating panel DFT system with a tube for the leek plant to grow in each plant hole was introduced. Faster expansion, sand-free high quality, but also higher investments and ongoing costs, were the results.

Figure 42: Lettuce grown in a deep flow pond.

Source: en.wikipedia.org/wiki/Deep_water_culture

***Concrete floor*:** A concrete floor and EF irrigation are included in one variation of the wide bed strategy. This kind of system is mostly used to root cuttings in 60 * 40 cm crates, raise young plants for fruiting vegetables, and grow labor-intensive pot and bedding plants. There could be heating tubing installed on the concrete floor. The floor is separated into portions for watering, and irrigation water enters each section through gutters that also serve as drainage systems. To aid drainage, the floor has a slope between 0.2

and 0.4 percent. The difference in water height becomes too significant to achieve a uniform water level for all containers on the floor if sections are too wide or if the slope is too steep. Depending on the size of the container, the water depth during irrigation can be as much as 5 cm. Containers positioned on an EF floor must have drainage systems. This kind of growth system is best suited to highly automated procedures in which containers are placed or retrieved using specialized machinery (automation/robotics). A concrete floor should last for the same amount of time as a greenhouse (around 15 years). It is difficult to determine whether using a concrete floor is economically feasible given that tearing it up is just as expensive as having it built in the first place. The fact that repeated watering with fertilizer solution would promote algal growth is also pertinent. Additionally, staff moving around on the floor while planting and removing plants could introduce infections into the irrigation systems. Special attention to hygiene is required, such as using boots exclusively in the greenhouse and washing the wheels of trolleys as they go from one compartment to another. Flexibility should be considered when designing a floor. For year-round lettuce cultivation, Birch (1979) created a concrete floor system with a 10 cm trough-profile. The design includes transport equipment. Unfortunately, such a system forbids switching to any other crop, forcing the producer to realize the investment's payback without having any options for altering the cropping system over time.

Above-Ground Production Systems

It has also been discovered that raising the production system above the ground can provide a number of advantages, including improved air circulation around the plants, a decrease in the occurrence of plant diseases, and increased worker productivity. Some plants also gain from better control over the root zone's temperature. Crops can also be raised above ground level to facilitate harvesting by giving workers easier access to plants or fruits, which will enhance working conditions in the greenhouse. Raised systems are used for crops such as potted plants, strawberries, lettuce, roses, tomatoes, peppers, cucumbers, and gerberas (*Gerbera hybrida*). They can be either stationary or mobile. Crops can be planted in single rows or in beds, just like production systems that are positioned on the ground. The majority of container-grown blooming potted plant production is produced in above-ground beds, sometimes known as "benches."

Single rows

To create a single or double row of container-grown plants from which harvests are regularly gathered, racks made of metal or wood are put on stone blocks or risers made of metal or treated lumber. Either a row of pots or a series of long pans or troughs might constitute the root zone. Roses, gerberas, and strawberries are a few examples of crops that have been cultivated in such systems. Farmers are now testing a number of other crops. Depending on how much the ground can sustain the system without subsiding, the racks may be partially or fully embedded in the ground. A metal rack with containers with a diameter of 15 to 25 cm is one example of this kind of system, and below the containers is a trough that collects drainage water. This trough may be constructed from coated metal, polypropylene, or PVC. On-site customization of the system for the grower is possible. Another type holds bags or slabs on long trays that are 100–133 cm long, 15-20 cm wide, and 7.5–10 cm high. At the intersection of two trays, a rack supports each tray separately (every 100-133 cm). In more advanced systems, containers are hung inside an aluminum frame. A rack with adjustable height is used to hold two frames. A one-row method, in which the supporting structure is firmly set in the subsoil, can be used to cultivate crops outside. Additionally, there are specialized parts that may be used to fit a rain cover over the one-row system.

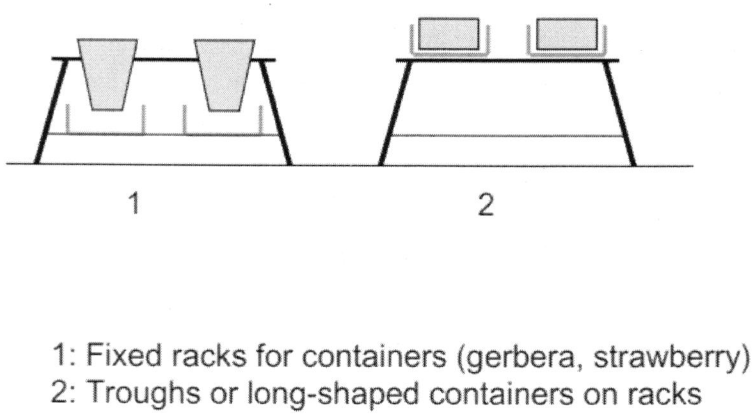

Figure 43: Example of an above ground system.

Source: Van et al., (2019).

Suspended troughs

The development of some systems has also resulted in the employment of steel cables to hang the complete production system from the frame of the greenhouse. There is no requirement for racks to support the system from below in this instance. It is obvious that such an installation is only possible if the greenhouse is built to hold this substantial weight in addition to the greenhouse covering material. The suspended rows have a smooth slope of 0.2 percent to 0.5 percent thanks to cables that are set at intervals of 2.5 to 4 meters and attached to the greenhouse's structural supports. For this purpose, troughs with different cross-sectional forms are available. The "plateau trough" (10–30 cm wide) is a regularly used design where slabs or bags are placed on a nearly flat plateau with short channels (4.5 cm) on either side (approximately 10 cm lower) for drainage water collecting. Particularly, it is helpful for preventing root growth in the drain water gutter is the space between the bottom of the bag and the water.

A-frames

It has been examined whether using A-frames or arches can improve the effectiveness of space usage. While Leoni et al. (1994) discussed the use of an A-frame for the development of tomato and lettuce with aeroponic watering, Morgan (1985) established the technique for a trough system with 12 layers. The dimensions of Morgan's system were 240 cm at the base, 40 cm at the top, and 230 cm high. It was created for a 3.2 m greenhouse span and could hold 12 lettuce troughs with an NFT system. With eight to nine crops every year, this design increased the cropping surface (up to 40 lettuce plants per m2) (Morgan, 1985). The Leoni et al. (1994) device was made up of two 100 cm broad polystyrene panels. The base of the triangular A-frame produced by the two panels measured 120 cm. The system had two A-frames, each measuring 3.2 meters. The lower plants in these A-frame systems are less productive because they receive less light than the higher plants do. In the technique devised by Leoni et al. (1994), there is also a chance that plants will develop a lettuce head that is crooked. Such systems are routinely displayed at demonstration locations, although they are not currently used in commerce.

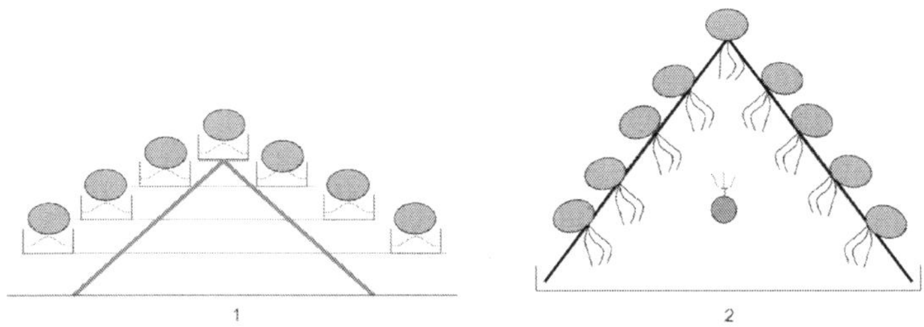

Figure 44: A-Frame systems (1. Troughs with NFT channels 2. Polystyrene panels with aeroponic misters.)

Source: Morgan (1985): Leoni et al. (1994).

Benches

In order to reduce the amount of aisle space in the production area, benches are typically utilized to store many plants side by side. The width of the bench is determined by how far workers must reach into the crop to control the plants. Fixed, mobile, or portable benches are all types of benches. In the greenhouse, fixed benches are in a fixed position and cannot be moved. When aisles are required for crop upkeep, they are primarily employed for the cultivation of potted plants. Space usage is not very high (65 percent -75 percent). The material used to construct rolling benches is lighter weight reinforced material, and they are positioned such that they sit on two long metal pipes that run the entire length of the bench. The bench may be moved side to side using cranks on the ends of the pipe. To keep a bench from rolling off its supports, the distance that it can move is carefully limited with stops. Since every bench in such systems may be moved, the area utilization is raised to between 85 and 93 percent, just one aisle is required per greenhouse span. The introduction of aluminum benches with a water-tight polyester contoured bottom for EF irrigation was another development step for rolling benches. Wider benches, some measuring 1.5 meters wide, have also become more common. However, it seems that this would be too big for rolling benches filled with labor-intensive crops. A bench width of less than 1.4 m is advised for ergonomic reasons because the typical Northern European worker's reach is 65–70 cm. The width of the bench would need to be reduced if the workers' average height was shorter.

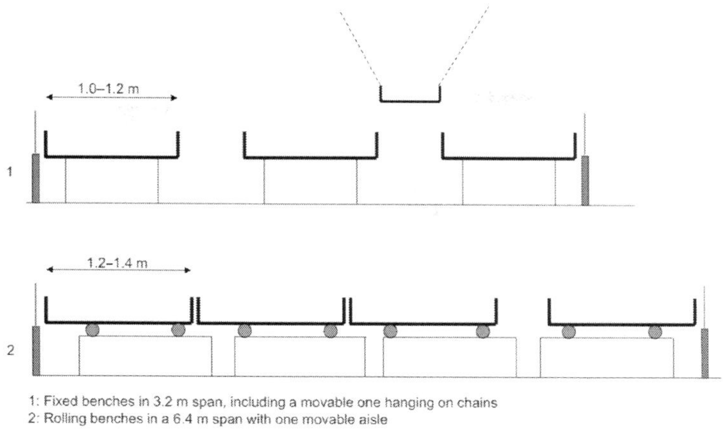

Figure 45: Fixed and rolling benches.

Source: Van et al., (2019).

***Movable benches*:** For usage with plants that need a lot of handling, mobile bench systems were created in areas with high labor and land expenses (spacing). By rolling the tray to a different place on this transport system, a special transport system is devised to allow a complete tray of plants to be relocated at once. Instead of the employees needing to spend time traveling to the plants, this enables the plants to be quickly transferred in huge batches to the worker. Aisle space can also be reduced, and in some cases eliminated, using such systems. Although movable benches can be designed specifically for the grower, they are often available in 3, 4, or 6 m lengths. The bench's breadth ranges from 120 to 150 cm. As a result, the greenhouse uses 90 to 95 percent of its available space. Many of these mobile benches also include a watertight bottom made of PVC or polyester, allowing for EF or capillary mat irrigation. For the creation of cut flowers, the idea of delivering the plants to the employees is now being investigated. Transporting the single rows of plants from the greenhouse to a separate harvesting area is being explored for the labor-intensive crops rose and gerbera, which require a daily harvest from the same plant. This arrangement makes it simple to maintain or harvest all of the plants in the row. With this kind of system, there are still two issues to be resolved: One is the creation of the best drip irrigation system that can be disconnected and attached quickly (and automatically); this is a technical barrier that can probably be overcome. The second issue has to do with how often such compact rows of plants need to be relocated and how quickly they can be transported. Transport of the rows must occur every 4 to 8 hours. Rows in one span enter the main path where maintenance or harvesting is done; it then exits via a different aisle to return to its original location. It might be challenging to control the cycles of harvest for each plant. The largest problem still seems to be economic in nature, as there has been a minimal increase in output but a significant increase in operating expenses.

Figure 46: Movable benches.

Source: www.mobilegrowsystem.com/

***Individual plant*:** Plants and crops are being cared for in a more customized way in intensive greenhouse farming. The smallest movable unit for rolling benches is a 1.5 m wide bench with a 20–50 m total span length. The bench measuring 1.5 m by 3-8 m is the smallest transportable unit for benches. One row of plants measuring 0.8 by 3–8 m is the smallest mobile unit for movable troughs. In the end, it's possible that specific plants will be relocated and handled differently. The most personalized care is provided through the "walking plant system." They are put on an adjustable belt after being potted up in containers. The plants are loaded onto a moving belt at the bottom of a trough by a machine. The gardener programs this system to arrange the plants in a specific row with specific spacing. By employing one automation system or machine to remove the plants from the trough and transfer them onto a primary transport belt, and a second automation system to set the plants into a trough at specific spacing, additional machines are employed to position the plants within the trough. Recently, equipment with cameras for grading, disease scouting, and spraying became accessible

CHAPTER 7

SPECIFIC SOILLESS CROP PRODUCTION SYSTEMS

CONTENTS

OVERVIEW

The number of soilless growth techniques is growing. A sizable proportion of the many various crops that have been attempted have been adopted by growers. A number of these manufacturing systems will be discussed in this section.

LEAFY CROP PRODUCTION IN SMALL-SCALE SOILLESS SYSTEMS

Food production in homes and small businesses can take many different shapes. The majority of people have traditional home gardens that use local soil, although interest in producing vegetables is not restricted to those who have adequate outdoor and in-ground locations. Many times, a gardener might not have access to a piece of land or the soil might be so bad that growing things in the ground is not an option. Many people have the choice of soilless farming or hydroponics, which makes it possible to grow vegetables on a small scale where conventional gardens are impractical. Those living in urban areas with limited space, a sunny patio, or a variety of different locations and scenarios can benefit from growing their own food thanks to the growth systems and methods used in soilless agriculture. Although it has been possible to grow plants without soil for a long time, opportunities for both commercial growers and home gardeners are growing as a result of these approaches' growing scientific understanding and practical use. In an effort to educate gardeners, students, youth, and other non-commercial producers about these hydroponic practices and techniques, the University of Tennessee Extension and University of Florida Cooperative Extension systems have collaborated to create this educational pamphlet.

Growing Systems

For green crops, there are numerous options for soilless growing systems. Make sure the system you choose is appropriate for the site, the amount of time you have, and your desired crops. The two most popular growth techniques for leafy crops are recirculating and floating systems. Vertical, or tower, systems, as well as recirculating nutrient film technique (NFT) systems, are all easily accessible. Additionally, a variety of sizes of floating or deep-water systems can be purchased or built to suit different purposes, locations, and budgets. Although there are more choices, when solution conditions (temperature, oxygen, electrical conductivity [EC], alkalinity/

acidity [pH]) are carefully managed, these two growing methods promote rapid crop growth and very simple management for leafy crops. On the market are a variety of full systems, often known as hobby hydroponic kits. Small full systems have advantages like easy setup and customer support options. Nevertheless, the cost is typically higher than DIY systems. Other crucial factors include location, production potential, management requirements, food-safe materials, and system versatility, in addition to cost and building or assembly time. The advantages of particular systems and vendors won't be the main topic of this talk. Instead, the fundamentals of managing nutrient solutions and growing green crops are the main focus because these techniques are essential regardless of the production system selected.

Figure 47: A low-cost floating hydroponic system.

Source: www.angi.com/articles/

Location

Prior to making a purchase, it is crucial to decide where your growing system will be installed. In the backyard, educational, or little-used commercial greenhouses, several small-scale hydroponic or these soilless systems can yield successfully. Additionally, some systems that are more weather resistant can be used for outdoor production on porches or patios. No matter the region, most vegetable crops require six to eight hours of natural light per day, whereas fruiting crops benefit from more than eight hours. Leafy crops can be cultivated in slightly lower light conditions, but it may take them longer to mature and may cause the plants to sag or have poor color. In order to support photosynthesis and lower the risk of leaf disease infection, air movement is also necessary. Make sure there isn't a lot of heat accumulation or reflected light at the location since this could stress the plants and lower yield. Depending on the place and season, a tiny greenhouse or outdoor space may not be ideal. For instance, a totally exposed porch in Tennessee in the middle to late spring can yield a fantastic harvest of lettuce, whereas Florida might require some partial shade at the same time of year. With the aid of extra lighting, soilless and hydroponic cultivation can also be carried out indoors. However, when lighting and ventilation systems are required, production costs increase. It will also require equipment, as well as knowledge of the required light levels and time. Sites with natural light will be the main topic of our talk.

Production Potentials

Consider the quantity of green vegetable crops required in proportion to time, expense, and area when examining systems. In ideal circumstances, soilless systems may produce lettuce and many other leaf crops in six to eight weeks (two to three weeks in transplant stage and four to six afterwards). Harvests with mixed or young leaves might be harvested earlier than mature crops. Although excessive light and warmth can significantly lengthen production times, productivity predictions can be helpful in determining system needs and crop dates. Let's use a hydroponic system with 36 plant spaces as an example, where a lettuce crop is being grown. Lettuce takes four weeks to mature after being transplanted. A grower or gardener would frequently have plants of various ages developing to offer consistent harvest potential. Under ideal conditions, eight heads of lettuce would be ready for harvest each week if young lettuce were transplanted into a fourth of the system each week. Using the same staggered planting strategy, smaller systems with six to eight spots in total might only yield one to two heads per week.

Ease of Management

It can be good to think about how nutrient solutions are managed and how the system is cleaned. Management ease is not always easy to foresee. The accessibility of accessing this solution for testing, pH or fertilizer correction, or solution changes is significant since controlling pH and fertilizer concentrations in the nutrient solution is crucial. Also take into account how simple it is to change the nutrient solution and clean the system. For instance, whereas it takes much longer to replace a constructed floating bed with a plastic liner, a tiny recirculating system tank can be cleaned and emptied in a matter of minutes. It is recommended to regularly clean systems to avoid the formation of algae and pathogens as well as to combine fresh nutrient solutions to avoid nutrient imbalances.

Common Leafy Crops

Types of head lettuce

Because it requires more space, more time, and rather precise conditions to mature, iceberg lettuce is not a typical hydroponic crop. Butterhead or bibb lettuce is the most popular green crop cultivated in most soilless systems. Although bibb lettuce's softer leaves are less popular in field lettuce production and do not ship as well as iceberg and romaine, the flavor is frequently favored for home production. Another explanation could be that bibb lettuce has been the focus of most lettuce breeding for hydroponics and greenhouses, leading to a number of reliable cultivars with strong disease resistance. Although soilless production of romaine lettuce is growing in popularity, it can be difficult because to a higher risk of quality problems like tipburn (death of young leaf edges). More open heads are common in romaine cultivars grown in hydroponic systems than in the common romaine hearts found in supermarket stores. Because oakleaf and other leaf lettuce varieties are less prone to tip-burn and yield an appealing, largely reliable, and timely crop, they can be a suitable choice. Leaf lettuce cultivars can also be developed to a higher level of maturity. You can also produce smaller, serrated lettuce varieties like lolla rossa and green leaf lettuce. Although the lolla rossa type lettuces normally grow more slowly and frequently yield less overall than other varieties of head lettuce, they do have advantages in terms of unusual leaf forms and colors.

Figure 48: Boston Bibb Lettuce.

Source: specialtyproduce.com/

Mixed Lettuce

Baby lettuce is just lettuce that has been harvested when it is still young. For immature harvest, a variety of romaine, leaf, and oakleaf cultivars can be effectively cultivated. Multiple seeds are planted in one growing space when plants are young to encourage rapid growth. In essence, these mixed cubes are a mixed bag salad that was grown and harvested at the same time. A tasty and aesthetically pleasing crop that can be harvested in a variety of sizes is produced by the possibilities for blending colors and leaf forms.

Figure 49: Mixed Lettuce.

Source: chefs-garden.com/blog/

Basil

Basil is the second most popular leafy crop cultivated in hydroponic systems after lettuce. Basil is frequently picked more than once, unlike lettuce, which is typically harvested only once when it reaches maturity. Harvesting growing points and young leaves promotes branching and growth, which in turn produces more harvestable leaves and sensitive stems. Basil plants are frequently able to be grown and harvested for several weeks or months. Basil is a warm-season crop; therefore it prefers higher light levels and temperatures than are normal for growing lettuce. Basil can actually be cultivated in systems for fruiting crops with fertilizer amounts that are comparable to those for tomatoes and cucumbers.

Figure 50: How to Grow Hydroponic Basil.

Source: ponicsarea.com/grow-hydroponic-basil

Kale and other Brassica crops

Other leafy plants that thrive in hydroponic systems include kale, mustard, pak choy, and mizuna. Many of the crops and cultivars utilized in hydroponic systems were really intended for use in soil, despite the fact that many lettuce varieties were bred and developed for soilless production. Because certain crops and cultivars do not stay as compact as needed and may need to be harvested at earlier stages, sampling a variety of crops and cultivars is one of the best ways to find out what works well in the site and system.

Figure 51: Hydroponic Kales.

Source: news.cornell.edu/

The Basics of Nutrient Solutions

Source of water

Both well water and city water can be successfully used, but it is better to send a water sample to a public or private lab to find out how much salt, chloride, and other micro- and macronutrients are present. Because there is no soil to serve as a buffer and the plant roots are frequently directly submerged in the solution, poor water quality can be more harmful when using hydroponics. Due to salt stress, high levels of sodium and chloride might be problematic (75 parts per million is often used as a general upper limit). Because water softeners remove minerals that lead to hardness (such as calcium and magnesium) and replace them with sodium, you should avoid using water that has been through one. Reverse osmosis treatment or dilution with water from another source may be necessary for a source of water with high salt content. For soilless systems, nutrient solutions are frequently based on the source water, which may contain a variety of dissolved minerals. Although several of these minerals are plant nutrients, they may be present in higher concentrations than what plants need. High concentrations of particular minerals, such as calcium and carbonate, can cause problems with solution formulation, pH control, and plant nutrient uptake.

Fertilizer materials

Because nutrients differ depending on the growing system, only use those made for hydroponic systems. For instance, a hydroponic fertilizer may contain more calcium and magnesium than fertilizers used with soil or potted greenhouse crops. Additionally, compared to fertilizers for soil-based cultivation, the majority of hydroponic fertilizers typically contain all micronutrients. The use of greenhouse-grade fertilizers is particularly crucial since they are totally soluble and do not contain filler substances typical of fertilizers intended for soil, which can block pumps and irrigation systems. Fertilizers for hydroponic plants often come in liquid or dry/granular form. The most economical fertilizers are frequently the dry ones. They can be purchased as individual fertilizers that are combined in specified ratios dependent on source water, or as a premixed solution. Liquid fertilizers are frequently premixed and not water source-specific. Although they are easy to employ in small hydroponic systems, they are challenging to modify for unique source water conditions. A custom mix is an option if your water is of poor quality, such as having a high pH or carbonates, however, such mixes are frequently unnecessary on a small scale. Despite the potential cost savings of making your own fertilizer, most small-scale farmers choose premixed fertilizers since they are more practical.

Managing Nutrient Solutions

Fertilizer use in soilless systems: There are two methods that dry or granular fertilizers can be applied in soilless systems. To make a nutrient solution that plants can directly utilize, they can be added in known quantities to a known volume of water. This procedure is comparable to mixing soluble fertilizer like Miracle-Gro with water (which is NOT advised for hydroponic systems alone). This is what we commonly call a feed strength solution. Concentrated solutions can also be blended using fertilizers. It is essentially making a mixture like the liquid fertilizers outlined below when dry or granular fertilizers are used to make liquid fertilizer concentrates. In addition to being more accurate than applying dry fertilizers directly, combining concentrates that may be used repeatedly can save time and money (liquid volumes can be measured more quickly and accurately than weights). Both methods are covered in the instructions for many fertilizers. In the hydroponics industry, pre-mixed liquid fertilizers are quite popular. They are added to source water in exact proportions and contain dissolved minerals. Instructions can be given in ounces per gallon, which is the volume of fertilizer liquid added in relation to the volume of source water. Dilution ratios can also be used to combine liquid fertilizers. An illustration would be a concentrate to water or a dilution ratio of 1:100. Accordingly, 100 gallons of nutrient solution would be produced when 1 gallon of fertilizer concentrate was mixed with 99 gallons of source water.

Managing electrical conductivity (EC): Small-scale hydroponic farmers frequently mix nutrient solutions in accordance with fertilizer label instructions without having a way to check the final solution's concentration. This method may be effective but bear in mind that fertilizer recommendations are based on typical crops and water sources. Mixtures can be far from optimal if the water already contains a lot of dissolved minerals. Therefore, having the ability to assess the concentration of all minerals and nutrients in solution might be highly valuable. There are two typical approaches. To start, samples can be examined in a lab to find out how much of each nutrient is present in the solution. Sending solution samples to a lab on a regular basis can be a desirable practice, but the time and cost involved typically make this impractical. Measuring electrical conductivity (EC) is the second popular technique for estimating the total number of dissolved ions in solutions (from fertilizers and source water). Measuring EC is an easy way to check that nutrient solutions are mixed properly and gives a clear indication of the overall strength of the fertilizer. You can buy portable instruments to measure the EC in solutions. Frequently, pH probes are added to these units to create multipurpose handheld equipment. Fertilizers can be blended to target ECs that offer particular nutrients based on the ratio of nutrients in the fertilizer with the right testing and validation. Therefore, although while EC cannot precisely tell us the level of a specific nutrient at a given time, it can be utilized to avoid significant mixing mistakes and direct management of solution strength during growth. For instance, the majority of leafy crops are grown with EC levels of 1.5 to 2.0 milli-Siemens per centimeter in the fertilizer solution. EC levels may decrease as a result of the plant absorbing nutrients during growth. This would indicate that additional fertilizer was needed to replenish the nutrients the plant had absorbed. Up until the EC once more reached the target level, these fertilizer materials would be added to the reservoir. The EC in the reservoir might increase if warm weather allowed the plant to transpire a lot of water. To bring the EC back down to the desired level, the grower could add source water. These instances show how crop growth can be controlled with EC. The concentration of all dissolved nutrients in a solution is determined by a property called electrical conductivity. Additionally, it is possible for the EC value to be steady over time even if particular amounts of important nutrients (most frequently nitrogen and potassium) are not met. The most typical method is a change in the volume of the majority of the nutrient solution in the reservoir tank in the absence of equipment to measure individual nutrients (often referred to as a tank change). The optimal nutritional balance will be restored, promoting crop development and productivity, when nutrients and fresh source water are combined. Crop, season, and container size all affect how frequently nutrient tanks need to be changed. For many actively developing crops, it is typical to switch the solution every one to two weeks during the primary growing season.

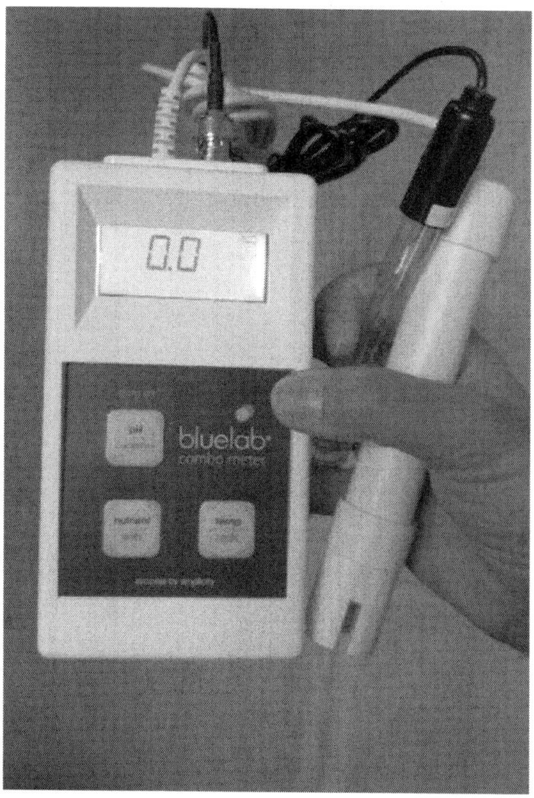

Figure 52: Hand-held pH and EC meter.

Source: Singh & Dunn, (2016).

Assessing pH in soilless systems

A measurement used to determine a solution's acidity or alkalinity is the pH. It is possible for plants to use nutrients most effectively by maintaining proper pH levels. Successful soilless production depends on accurate pH measurement and management. There are two popular ways to check pH:

- portable pH meter. These tiny electronic meters have electrodes that can quickly and accurately determine a solution's pH. (Figure 52). They can be a bit pricey ($80 and over), though, and are only accurate when properly calibrated.
- solutions for colorimetric (color-changing) indicators. You can buy dyes that change to specific colors at particular pH levels. A quick and low-cost method of determining pH can be to add a few drops of these indicator solutions to a sample of nutrient solution and compare the color to a pH chart (if you are not red-green color blind). Another colorimetric method that can be helpful for novices to accurately and affordably measure pH is litmus paper. Although it might not be as accurate, it can be helpful for tiny hobby systems with few tools.

Managing pH in soilless leafy crop nutrient solutions

The pH is maintained in a range (often 5.5 to 6.0 pH) that will keep nutrients in their most absorbable forms for plants. A proper pH balance will also prevent nutrients from precipitating, blocking pipes, or accumulating on pumps. To achieve the desired pH, basic or acidic elements may need to be added, depending on the source water. To raise and lower pH, use diluted acid or base solutions. Initial pH adjustments made while combining fertilizer solutions are typically insufficient to maintain ideal pH throughout the plant's growth. Since nutrient uptake by plants and the reaction of solutions with air may all quickly affect pH, regular attention is required to maintain pH levels. The buffer capacity, or ability to withstand pH variations, varies depending on the dissolved components in the source water. When exposed to acids or bases that modify pH, some source waters may react quickly while others may react slowly. In order to avoid overshooting the objective, it is crucial to add acidic or basic compounds slowly when altering pH. Hydroponic system pH adjustments will get easier to approximate with time and practice.

Growing Transplants

Substrates

Before being planted in the growing system, most green crops are grown for 10 to 20 days in cubes or plugs. Since the early stages of growth consume significant space in the main system, this is the strategy that uses the least amount of space. Additionally, smaller growth systems can be maintained to facilitate germination more effectively than larger ones. For soilless systems, substrates are produced to be consistent and simple to use. A substrate is just the substance where plant roots reside. To hold the seed firmly during germination, for instance, each cube will include a dibble (indentation in the surface of the cube). Additionally, this indentation increases the moisture and humidity levels directly around the seed, lowering the risk of the seed drying out. As is typical in soil or greenhouse mixes, the majority of seeds are not covered. Roots can obtain moisture, nutrients, and air in the right amounts thanks to the substrate. Some are created using natural materials like peat and coconut husks (coir). To help the cube maintain its shape, several of these materials contain a polymer binder. Others might be protected from loose particles in both recirculating and floating systems by a tiny mesh or fabric shell. Other substrates are comprised of inorganic components or synthetic polymers. Rockwool (made from basalt rock that has been extremely heated) and foam cubes are the two most popular inorganic materials. Ideal air pore diameters can be created in both of these materials by engineering. Additionally, these materials are available in a variety of sizes to accommodate various growing methods. Larger cubes can be utilized to develop a larger transplant and may offer additional protection from changing solution conditions, albeit being generally more expensive.

Figure 53: Foam cubes with indent for lettuce seed.

Source: CropKing Inc.

Seeds and seeding

You can buy seeds either raw or pelleted. Although pelleted seeds cost more, they can be simpler to handle and seed by hand. Since manual sowing is still the norm for small-scale farmers, they can be beneficial. Use what you buy within 6–12 months to lessen the chance of reduced germination. In solid seeding trays, cubes or plugs are typically put for germination so that water solution can be delivered from the bottom. It is ideal to seed into plugs or cubes that are uniformly wet. You can use pH-balanced pure water or diluted nutrient solution for the first several days after seeding. By 10 to 15 percent overseeding, you can make up for uneven or unsuccessful germination. Additionally, this approach guarantees that only the most consistent seedlings are transplanted. For head lettuce, put one seed per cube; for mixed lettuce, use several seeds per cube. It might be necessary to use raw seed if you want baby lettuce that has numerous (more than three to four) plants per cube because several pelleted seeds quickly fill up the dibble in the cubes.

Figure 54: Young lettuce seedlings.

Source: CropKing Inc.

Germination conditions

Make sure the cubes are evenly moist but not saturated after adding the seeds. In seedling trays (with solid bottoms) apart from the growing system, transplants can be initiated. In most cases, 1/4 inch of water in the tray is adequate. Applying pure water after sowing can be okay for the first few days, but using fertilizer solution as soon as germination happens is crucial. The pH levels of this water or solution will need to be monitored frequently because they can fluctuate quickly and affect seedling growth. It is also possible to carry out germination and seedling production in a nursery space that is connected to the growing system. The solution no longer needs to be managed in a separate tray thanks to this technique, but it's crucial that the right temperatures can be maintained. The ideal germination temperature for lettuce and other green crops is between 70 and 75 F. Observe advice specific to the crop and cultivar you are utilizing. If necessary, heat mats can be utilized to raise the temperature. Also bear in mind that germination may be hindered by hot temperatures. Above 85 F, the germination of cool-season crops like lettuce, spinach, mustard, and kale can be reduced. Avoid exposing trays or pads to direct sunshine during unusually warm months to help slow down temperature increases that could harm seeds or seedlings. During warm weather, it may be ideal to germinate seeds indoors or in a shady place; however, once germination has started, remove the seeds from the shade. Usually, germination will start between 24 to 48 hours. Raw seeds may germinate a little bit quicker than pelleted seeds since the coating on the latter needs water to be broken down or cracked. It is best to add a diluted nutrition solution after seed leaves (cotyledons) start to emerge. For the first few days, seedlings are frequently begun on a half-strength solution (0.5-1.0 mS/cm), and by the time transplanting takes place, the strength of the nutritional solution has been gradually increased to match the main system.

Transplanting and Growing Out

Transplanting

Crop, environment, and cube size all affect how long passes between sowing and transplanting. Transplanting frequently happens 12 to 18 days after sowing. After developing three to five genuine leaves (excluding the cotyledons) and roots that are visible on the bottom of the cube, seedlings are deemed mature enough for transfer into a hydroponic system. These roots can aid the young plant in adapting after being transplanted, whether in a recirculating or floating system. The seedlings will start to stretch and these elongated stems are easier to break during transplanting, thus it is preferable not to wait until crowding begins. Separating each individual cube before putting it in the production system is known as transplanting. Generally speaking, employing larger growing cubes will allow the plant to be more mature when transplanted. Make sure the cube itself will be in contact with the nutrition solution while taking care to avoid damaging the young stem or roots.

Environmental conditions for leafy crop growth

The ideal daytime temperature range for lettuce and other cool-season green crops is 65 to 75 degrees Fahrenheit. At lower temperatures, they will still grow, only more slowly. Conversely, as temps hit the mid-80s, growth will also be slowed. High temperatures can also lead to quality problems like bolting or tip-burn, a physiological condition connected to poor calcium uptake in quickly developing leaves (stem elongation and the beginning of seed production). In many regions of the United States, spring and fall offer the best lighting conditions for growing lettuce and many other cool-season leafy greens. In the Mid-South and occasionally even in Florida, winter light levels are frequently insufficient for the best development and quality, whereas, throughout the Southeast, summer light and heat may be too intense. Shading can be useful in the heat (and sometimes late spring and early fall). Wintertime may benefit from more lighting, but it might be expensive. Remember that taking extra measures to manage the nutrient solution or environmental factors (such as lighting) can frequently boost growth, production, and quality.

Harvest and Storage

Harvest procedures

The crop, the growing environment, and individual tastes all influence when to harvest. Bibb lettuce can often be harvested when it weighs between 5 and 8 ounces, however, this is only a general guideline. Oakleaf lettuce often grows at a comparable rate and at the same time. Romaine could grow more slowly, and mixed leaf crops are typically picked when they are younger. Kale, mustard, and the other greens all grow at different rates according to cultivar and taste. By harvesting the leaves rather than the entire plant as is done with lettuce, it is possible to harvest kale, basil, Swiss chard, and several other leafy greens more than once. To lessen stress and water loss on plants after harvest, whole plants or leaves should be harvested during cooler parts of the day. The plants will still be turgid (have high-water content in cells) from the previous night, thus early in the morning is frequently best. The channels, boards, or other growing systems are used to pull the plants. Plants can either have the cube left connected or the entire cube and roots taken off at the base. Roots can be looped around the cube or trimmed at the bottom. To prolong storage life, remove any yellow, damaged, or diseased leaves.

Storage and quality

Crop health and quality at harvest, as well as handling and storage circumstances, all have an impact on the crops' storage durability and quality. Healthy plants can be kept in good quality for many days, and even up to weeks for some crops, if they are harvested in good shape, chilled rapidly, and stored quickly. Although they might be pricey in small quantities, "crispers" or "clam shells" made of plastic are a good way to protect and store lettuce. Plastic bags can also be useful, although when stacked, injuries are more common. Bags may also speed up the drying process for crops or keep moisture from coming into contact with the leaves and stems. The best conditions for storing lettuce are between 35°F and 38°F with a high humidity level to minimize water loss. The ideal storage temperatures will be similar for many other leafy crops. Basil is an exception, though, as it is delicate to chilling and can soon decay unless stored above 50 F.

TOMATOES, PEPPERS AND CUCUMBERS IN SMALL-SCALE SOILLESS SYSTEMS

Food production in homes and small businesses can take many different shapes. The majority of people have traditional home gardens that use local soil, although interest in producing vegetables is not restricted to those who have adequate outdoor and in-ground locations. Many times, a gardener might not have access to a piece of land or the soil might be so bad that growing things in the ground is not an option. Many people have the choice of soilless farming or hydroponics, which makes it possible to grow vegetables on a small scale where conventional gardens are impractical. Those living in urban areas with limited space, a sunny patio, or a variety of different locations and scenarios can benefit from growing their own food thanks to the growth systems and methods used in soilless agriculture. Although it has been possible to grow plants without soil for a long time, opportunities for both commercial growers and home gardeners are growing as a result of growing scientific understanding and practical use of such approaches. In an effort to educate gardeners, students, youth, and other non-commercial producers about these hydroponic practices and techniques, the University of Tennessee Extension and University of Florida Cooperative Extension systems have collaborated to create an educational pamphlet.

Growing Systems

Fruiting plants cultivated hydroponically can be grown in systems that have been built by the grower themselves or in complete, ready-to-use kits that can be purchased. For small-scale soilless cultivation, these multiple or single self-contained bucket, bag, and slab systems are more desirable. The simplicity of installation and the opportunity for customer assistance are advantages of tiny full systems. The cost is typically higher than buying inexpensive parts and building a modest system, though. Take into account production potential, managerial needs, and versatility in addition to cost and building time. The ease with which nutritional solutions can be accessed, tracked, modified, and changed is another area of concern. Consider how simple it will be to clean the system in between crops as well. It is also crucial to consider the dependability of the system's components (such as the pumps, lines, etc.), as well as how simple it is to repair them and whether spare parts are readily available. Since buckets are more expensive to buy initially but will endure for numerous crops, reuse of system components is also a factor. Although cheaper in price, plastic bags or slabs will probably need to be replaced after one or two crops.

Figure 55: Dutch Bucket Hydroponics for Tomatoes.

Source: cropking.com/blog/

For leafy crops grown for a shorter period of time, nutrient solution recirculation is typical; these systems are occasionally referred to as closed. Open systems, also known as feed-to-drain systems, are frequently used in commercial hydroponics for fruiting crops because they make it easier to provide the ideal nutrient levels and reduce the danger of pathogens by only using the nutrient solution once. On a smaller scale, closed systems can be employed for fruiting crops, which will cut down on solution use and make the systems simpler to manage in a variety of settings. When hydroponic gardeners try to establish and develop larger fruiting crops in systems best suited for leafy crops, such as nutrient film and floating beds, they face some of the toughest difficulties. In terms of space for root development, plant support, and fertilizer management, these systems might be difficult. Tower systems can be used to grow some compact tomato or pepper plants, but keep in mind that these crops have different light and nutritional needs from the leafy plants in the system. Fruiting crops are also more susceptible to root diseases brought on by recirculating fertilizer solutions because they are grown for longer seasons than green crops. Beyond these warnings, this article does not discuss the advantages of certain systems or vendors. Instead, the fundamentals of managing nutrient solutions and growing fruiting crops are highlighted, as these techniques are crucial regardless of the production method selected.

Production Potential

The various fruiting crops have varying production cycles and yields. Tomatoes may be three to four months old (90 to 120 days) before their first fruit ripens, but if the plants are taken care of, harvesting can last for several months. After they begin bearing under ideal lighting conditions, soilless beefsteak tomato plants in commercial settings may produce a pound or two of harvested fruit per plant each week. Weekly tomato production over the winter will probably be less than 1 pound per plant due to the short days and limited light. In general, yields are lower since conditions are not always as consistently ideal. Despite often being picked for weeks rather than months, cucumbers reach their initial harvest within 30 to 50 days. Fruit per plant, as opposed to weight, is frequently used to describe cucumber yields. Depending on the variety, size, and time of year, a productive seedless cucumber plant might yield 15 to 60 fruits per plant. In recent years, smaller fruited cucumber types have grown increasingly popular. Each year, many crops of cucumbers are frequently grown. Cucumber varieties are actually created for various growing seasons. The size of the system that is required can be determined using broad yield estimations. A harvest of three or more pounds of beefsteak tomatoes each week, for instance, may be predicted for the late spring to mid-fall months if the hydroponic system had space for six plants. These values are simply guidelines because a variety of variables affect production. Over time, your own records will yield the most precise estimations.

Location

In backyard, educational, or little-used commercial greenhouses, several small-scale hydroponic or soilless systems can yield successful results.

Additionally, some methods that are more weather resistant can be used for outdoor production. The best natural light for fruiting crops, regardless of location, is eight hours or more each day. The right amount of light is essential for fruiting vegetable crops. Plants may continue to grow in inadequate light, although fruiting will be lessened. Air circulation is also required to boost photosynthesis and lower the risk of leaf disease infection, which is crucial for crops produced for an extended period of time. Make sure there isn't a lot of heat accumulation or reflected light at the location since this could stress the plants and lower yield. With the aid of extra lighting, soilless and hydroponic cultivation can also be carried out indoors. However, when lighting and ventilation systems are required, production costs increase. It also calls for familiarity with the equipment, duration, and light requirements. The conversation in this section is limited to outside locations.

Common Fruiting Crops

Tomatoes

The most widely cultivated soilless vegetable crop is the tomato, which comes in a variety of forms and cultivars. Although cherry, grape, plum, roma or paste tomatoes, as well as cluster or tomato-on-vine (TOV), which are clusters of numerous tomatoes still attached to the stems, are among the various varieties, beefsteak tomatoes are typically the most popular. There are tomato varieties made specifically for greenhouses. These cultivars are good options even when grown in a small greenhouse where production may not be the primary criterion because they are more resistant to diseases typical of greenhouses, like leaf mold and powdery mildew. Disease resistances in greenhouse tomato cultivars might not be as helpful if tomatoes are grown outdoors using a soilless system. Additionally crucial are taste, productivity, and growing habits. The majority of greenhouse cultivars are indeterminate, meaning that they continue to grow vertically and produce leaves in addition to fruit. They were created to bear regularly for several months. Numerous backyard gardeners and industrial field farmers cultivate determinate cultivars, which bear fruit over a shorter period of time and don't keep on producing new stems and leaves in addition to fruit for the entire crop. Hydroponic systems can be used to grow both determinate and indeterminate crops, although there are differences in the harvest times and plant care. Here, indeterminate plants are highlighted. The two cropping calendars for hydroponic, indeterminate greenhouse tomatoes are distinct.

According to the more northern timeline, transplants are sown in December or January, allowing for a mid- to late spring harvest. The crop continues to produce fruit into the fall after being harvested during the summer. When to stop picking and take the crop out of production is frequently determined by disease pressure and low productivity. Around more southern regions, seeds are sown in August or thereabouts, and they begin to grow in the late fall. To prevent having plants in the greenhouse during the warmest portion of the summer, they are harvested during the winter and spring. Both cropping calendars are feasible in Tennessee. Although the fall to spring cycle is more typical in Florida, both locations may benefit from supplemental lighting to attain the best yield over the winter. The best tomatoes to grow in a soilless hydroponic system are briefly mentioned;

***Cherry Tomatoes*:** A hydroponic garden is perfect for the cherry tomato plant. Gardeners that grow the sweet million cherry tomato benefit from its prolific production of a small, delicious, and resistant-to-crack tomato. Although the tomato is slightly bigger, the sweet gold is not as productive as the sweet million. The sun gold cultivar has a delicious flavor and is orange in color. The sakura is a more recent variation. Although the sakura does not produce many tomatoes, the fruit is strong and firm and is long-lasting. Cherry tomatoes from Favorita are well shaped and have a great flavor. A more recent variety, Conchita has a solid fruit and a lengthy shelf life.

***Beefsteak Tomatoes*:** The beefsteak tomato is one of the most often used tomato varieties for hydroponic farming. Trust tomato plants offer a huge, uniformly-ripening tomato weighing 7 to 8 ounces, making them a high-yield crop. The jelona yellow is a very strong plant that produces a firm, yellow fruit weighing between 6 1/2 and 6 3/4 pounds. Large, 8 1/2-oz. crimson fruits are produced by matrix beefsteak tomatoes. A new kind that is becoming in popularity is the Geronimo. The Geronimo tomato, which weighs 8 1/2 ounces and is firm and tasty, thrives in a hydroponic setting. The blitz produces crops in the spring and fall and has a medium-sized tomato.

***Cluster Tomatoes*:** Tomatoes that develop on the vine in bunches or clusters are referred to as cluster tomatoes or truss tomatoes. Cluster tomatoes of several types thrive in hydroponic environments. The Clarance is a highly productive plant that yields a tasty, deep red, 5- to 6-oz. tomato. The tradario is a huge cluster tomato with a good shelf life that weighs 4 1/2 to 5 1/2 oz. A 5- to 5 3/4-oz. tomato can be found in Tricia cluster tomatoes. The ambience cultivar yields a medium-sized, red, flavorful cluster tomato.

Cocktail Tomatoes: Cocktail tomatoes grow in clusters and are marginally bigger than cherry tomatoes. A cluster of 10 deep red, 2 oz. tomatoes, produced by the red joy cocktail tomato plant is typical. The tomatoes from Red Delight have a long shelf life and are hardy. Cocktail-sized Flavorino tomatoes resemble roma tomatoes in size. They taste full-bodied and delicious. Each truss supports seven to nine flavorino tomato plants. The fruit of the picolino is small, tasty, and resistant to cracking. Tomatoes in the picolino kind retain their flavor nicely. The orangion is a diminutive cocktail tomato that is orange, and the goldino is a cocktail tomato that is yellow.

Cucumbers

A crop with a quick growth rate, cucumber can work well in small-scale soilless production systems. Cucumbers can be grown in small systems using the same fertilizer solution as tomatoes. If bees are present to perform pollination, soilless methods can be used to cultivate garden-type seeded cucumbers. However, greenhouse cucumber types don't need pollination because they are seedless and resistant to common diseases in greenhouses. In actuality, male blooms are uncommon on greenhouse cucumbers. Therefore, if these cultivars are grown outside with other varieties of cucumber, pollinators can pollinate seedless cucumber flowers, causing them to produce seeds. For outdoor production, seedless cucumbers are also available. These cucumbers can be produced in greenhouses, and home gardeners may find the seeds to be less expensive. Three to five crops of cucumbers are normally grown each year.

Peppers

Similar to how tomatoes are cultivated, bell peppers can also be grown in soilless systems. Because they are more protected from moisture extremes, insects, and diseases than they would be outside, growing peppers to full-colored fruit maturity may be simpler in greenhouses and hydroponic systems. Large bells, however, are difficult to manage since changes in light, temperature, and nutrients can lower productivity. Instead of producing full fruit when conditions are poor, peppers frequently drop blossoms or little fruit. Small bells and different kinds of hot peppers can be easier to handle and offer a more reliable crop.

Figure 56: Bell Peppers Growing with Hydroponics.

*Source: http://learnhydroponicfarming.blogspot.com/**Eggplant***

A desirable medium-length crop that bears fruit earlier than tomatoes and peppers is eggplant. While there are varieties of eggplant cultivated in greenhouses, a variety of varieties can also be grown at home. Reduced pest damage is one of the biggest benefits of growing eggplants in greenhouses (such as flea beetles). However, in a greenhouse setting, spider mites prefer the host plant eggplants.

Table 9: Fruiting crop table to estimate timing and crop calendars

	Tomato	**Cucumber**	**Pepper**
Number of crops per year and typical calendar	One Winter harvest — July seeding. August transplanting for fall-to-spring harvest. Summer harvest—December seeding, January transplanting for spring-to-fall harvest.	Three to five New cucumber seedlings are started when older plants begin to slow in production.	One Winter harvest — July seeding. August transplanting for fall-to-spring harvest. Summer harvest — December seeding. January transplanting for spring-to-fall harvest.
Time from seeding to transplant	2-4 weeks	2-3 weeks	6-8 weeks
Time from transplant to harvest	7-10 weeks	4-6 weeks	10-12 weeks

Source: Natalie B. & Robert H., (2019).

Nutrient Solution Basics

Source water

Both well water and city water can be successfully used, but it is better to send a sample to a public or private lab to check the pH, salt, chloride, and micro- and macronutrient levels. Due to the lack of soil to serve as a buffer and the fact that plant roots are frequently immediately bathed in the solution, problems with a low-quality water source might be more harmful in soilless systems. Due to salt stress, high levels of sodium and chloride might be problematic (75 parts per million is often used as an upper limit). The use of water that has been softened is not recommended since sodium ions are substituted for the hardness-causing minerals (such as calcium and magnesium) in water softeners. To lessen potential problems with high sodium levels, a water source with high salt levels might need to be processed (using reverse osmosis) or diluted with another source. For soilless systems, nutrient solutions are frequently based on the source water, which may contain a variety of dissolved minerals. Although several of these minerals are plant nutrients, they may be present in higher concentrations than what plants need. High carbonate concentrations can hinder plant uptake of nutrients or the formation of solutions.

Fertilizer materials

Because they differ depending on the growing system, only use fertilizers made for hydroponic systems. For instance, a hydroponic fertilizer may contain more calcium and magnesium than fertilizers used with soil or potted greenhouse crops. Additionally, compared to fertilizers for soil-based cultivation, the majority of hydroponic fertilizers typically contain all micronutrients. Use fertilizers made for greenhouses instead of fertilizers made for soil since they are totally soluble and don't contain filler elements that can clog irrigation and pump systems. Fertilizers for hydroponic plants are often either dry/granular or liquid in composition. The most economical fertilizers are frequently the dry ones. They can be purchased as individual fertilizers that are combined in specified ratios dependent on source water, or as a premixed solution. Liquid fertilizers are frequently premixed and not water source-specific. Although they are easy to employ in small hydroponic systems, they are challenging to modify for certain source water conditions. A custom mix is an option if your water is of poor quality, such as having a high pH or carbonates, however, such mixes are frequently unnecessary on a small scale. Despite the potential cost savings of making your own fertilizer, most small-scale farmers choose premixed fertilizers since they are more practical.

Managing Nutrient Solutions

Fertilizer use in soilless systems

There are two methods that dry or granular fertilizers can be applied in soilless systems. To make a nutrient solution that plants can directly utilize, they can be added in known quantities to a known volume of water. This is what we commonly call a feed strength solution. Concentrated solutions can also be blended using fertilizers. It is essentially making a mixture like the liquid fertilizers outlined below when dry or granular fertilizers are used to make liquid fertilizer concentrates. In addition to being more accurate than applying dry fertilizers directly, mixing concentrates that may be used as a source repeatedly can save time and money (liquid volumes can be measured more quickly and accurately than weights). Both methods are covered in many fertilizer instructions. The liquid fertilizers market for hydroponics is dominated by pre-mixed fertilizers. They are added to source water in exact proportions and contain dissolved minerals. Instructions can be given in ounces per gallon, which is the volume of fertilizer liquid added in relation to the volume of source water. Dilution ratios can also be used to combine liquid fertilizers. An illustration would be a concentrate to water: a dilution ratio of 1:100. Accordingly, 100 gallons of nutrient solution would be produced when 1 gallon of fertilizer concentrate was mixed with 99 gallons of source water. Remember that the consumption of fertilizer solution will be significantly higher for a single-use (or feed-to-drain) system used to produce fruiting crops than for a system used to grow leafy crops. It can be difficult to mix solutions at the right concentration to feed crops directly. It might be more precise and time-effective to blend fertilizer concentrates in batches that can be utilized for several days or weeks. The ratios of nutrients for green and fruiting crops frequently differ. In order to apply the proper nutrients or fertilizer concentrations for each crop, it is best to separate leafy and fruiting crops in growing systems.

Nutrient solution delivery

Fruiting crops frequently receive nutrient solutions in a different way than leafy crop systems. Irrigation is given using feed lines and drip stakes that are positioned close to the root system, whether growing in perlite in buckets, rockwool slabs, peat/coconut coir in slabs, or upright bags. Drip irrigation is not continuous; it is performed in cycles. Plants with an unbalanced ratio of vegetative to reproductive growth will grow roots that clog drain lines and will be exposed to an ongoing flow of nutrient solution. One of the most frequent errors made by inexperienced hydroponics growers is this one. As an illustration, it is typical to begin transplants on feeding cycles that run every 30 to 60 minutes in perlite systems. As the weather warms, cycles can be spaced closer together. However, if the system is not recirculating, frequent feed cycles utilize more feed solutions. These cycles are scheduled to promote optimum aeration, avoid waterlogging, and keep the substrate from drying out completely. The timing of watering cycles is influenced by the substrate's rate of drainage as well as the surrounding conditions. Warmer climates necessitate more regular watering, as do plants cultivated in perlite and clay pebbles as opposed to peat or coconut coir. Systems frequently contain timers that automatically switch on and off the pumps. The nutrient solution is frequently only utilized once in bigger systems before being wasted or retreated (feed-to-drain systems). To improve sanitation and nutrition, this is done. Fruiting crops can be produced with recycled solutions in modest systems. These compact systems combine a feed-to-drain system's timer and timed delivery with a recirculating lettuce system's reservoir. The drainage from the buckets or bags is collected and utilized again when the irrigation is given on predetermined cycles. To avoid nutrient imbalances and lower the risk of sickness, it is crucial to regularly monitor the pH and electrical conductivity (EC, which is the total concentration of dissolved elements in the solution) in these systems and to change the tanks on a regular basis (every one to two weeks).

Managing electrical conductivity (EC)

Household growers can buying a general nutrient solution for hydroponics or making their own, if they want to save more money. You can make your own from NPK fertilizer with macro-nutrients (copper, cobalt, boron, iron, zinc), calcium nitrate, and epsom salt (magnesium sulfate). The ingredients are easy to find. The ratios are 1 unit NPK fertilizer + 1 unit calcium nitrate + half a unit epsom salt. If you go with the manufactured option, you should also be aware that there are nutrient products for the two main stages: vegetative and bloom. Fruiting crops, like the subject we're discussing, go through the following stages: vegetative-only, flowering and fruit set, and fruit development phase. Commercial growers will use different nutrients of different strengths for each of these 3 stages. But that would be just too difficult for us. Fruiting crops require an EC between 2.0 and 3.5. For the vegetative stage, you can have an EC between 1.8 and 2.4 and for the fruiting stage it can be between 2.6 and 3.0. iHarvest® comes with an EC tester. This enables monitoring the nutrition of the solution. Eggplant grows great with other plants that grow well in these EC ranges, such as cucumbers, peppers, tomatoes and even cabbage. Your eggplants are likely to remain healthy at EC ranges below 2.5, but will not produce fruit as quickly. The following steps should be performed for EC management:

- Fill the nutrient tank with tap or filtered water and add fertilizer. Base quantity on the manufacturer recommendation.
- Calibrate the EC meter probe using the buffer solution.
- Make sure the nutrient solution is stirred up and allow the reading to stabilize, which may take a couple of minutes.
- If the reading is higher than the optimum level, dilute the solution by adding more water, then repeat step 3.
- If the reading is below the optimum level, add nutrient concentrate until the optimum level is reached by repeating step 3.
- Rinse the probe in tap water and store in probe-cleaning fluid.

Managing pH

If you are looking for a reliable and affordable pH meter to help you grow eggplant hydroponically, then the HM Digital PH-200 is a great option. This meter is easy to use and helps you to ensure that the pH level of your nutrient solution is in the optimal range for growing eggplant. The PH-200 has a resolution of 0.01 pH and a accuracy of +/-0.02 pH, so it can help you to make precise adjustments to your nutrient solution as needed. The following steps should be performed for pH management:

- Set the desired EC value of the solution.
- Calibrate the pH meter probe using the buffer solution. Make sure the nutrient solution is stirred up and allow the reading to stabilize, which may take a couple of minutes.
- If the pH reading is high, add phosphoric acid, citric acid, vinegar or pH down products slowly. Wait several minutes before adding more. Repeat until the pH reaches the optimum range.

- If the pH is low, add potassium hydroxide, potassium carbonate or a pH up product slowly. Repeat until the pH reaches the optimum range.
- Clean the probe and store in cleaning fluid.

Nutrient solution pH range for fruiting crops is typically 5.5 to 6.0 pH, which will keep nutrients in the most accessible forms for plant uptake. The pH of the solution may need to be adjusted by adding basic or acidic ingredients depending on the source water. You can buy diluted acid or base solutions and combine them to change the pH. In feed-to-drain systems, only a small pH adjustment is required when combining nutrient solutions; however, if a recirculating system is being used, regular attention is required to maintain pH and EC values. The plant's intake of nutrients and the reaction of the solution with air can both quickly alter pH. The buffer capacity, or ability to withstand pH variations, varies depending on the dissolved ions in the source water. When exposed to acids or bases that modify pH, some source waters may react quickly while others may react slowly. In order to avoid overshooting the objective, it is crucial to add acidic or basic compounds slowly when altering pH. You will gradually gain confidence in your ability to predict how much adjustment is required as well as how your water responds to pH changes.

Growing Transplants

Substrate

The majority of fruiting plants are often begun as cubes or plugs and cultivated for a few weeks before being put in a growth system to finish growing until they are fully mature. The growing cubes used for leafy crops are often larger than those constructed of peat, coir, rockwool, and foam. It is advised to use a cube that is at least 1.5 by 1.5 inches. As a second transplant stage, bigger blocks (3- or 4-inch cubes) can also be employed. Don't feel as though rockwool or foam blocks are necessary because transplants can also be created in a manner similar to how transplants would be established for the garden. Transplants can also be transplanted into substrates in buckets, bags, or slabs after being cultivated in a loose, pathogen-free, soilless mix (typical bedding plant materials) in a seedling tray.

Seeding and germination conditions

Larger than leafy crops, fruiting crops' seeds are typically sown into uniformly moist cubes or plugs that are then placed in solid seeding trays for germination. This makes it possible for early plants to receive water solution from the bottom. To account for different seed germination rates, an additional 10 to 15 percent of cubes should be sown. This also permits some selection during transplanting and makes it possible to only transplant the most uniform seedlings. Make sure the cubes are evenly moist but not saturated after adding the seeds. Most of the time, one-fourth inch of water in the tray is adequate. Applying pure water after sowing can be okay for the first few days, but using fertilizer solution as soon as germination happens is crucial. The nutrition solution won't be required for the first several days after seeding because seeds are still germination. But it's ideal to utilize water that falls within the right pH range. The pH of this water or solution needs to be tested periodically because it can fluctuate quickly and affect seedling growth. The best germination temperatures for tomatoes and other warm-season crops are between 75- and 85-degrees Fahrenheit. Considerations for the crop and cultivar you are using should be followed, and low, high, or fluctuating temperatures can significantly lower the germination rate. If necessary, heat mats can be used to raise the temperature during germination. Avoid exposing trays or pads to direct sunlight in warm seasons to avoid higher than ideal temperatures (greater than 90 F), which can impair germination. During warm weather, it may be ideal to germinate seeds indoors or in a shady place; however, once germination has started, remove the seeds from the shade. A diluted nutrition solution should be supplied following the completion of germination and the appearance of seed leaves (cotyledons). A half-strength solution (0.5–1.0 mS/cm) is frequently used to start seedlings for the first few days before being gradually increased to between 1.0 and 1.5 mS/cm at transplanting.

Transplanting and Growing Out

Transplanting

Transplanting may take place three or four weeks after seeding, though this can vary. With larger cubes, larger and older transplants are feasible (crop support may be needed for these larger cubes). The size of the plants used in soilless production is frequently comparable to those used in conventional gardens. The same kinds of stocky seedlings with vibrant green color and no signs of disease or insects should be sought out. Make sure to completely cover the transplant cube's surface as you place the young transplant in the substrate. To keep the young plant from drying out, drip irrigation must be installed as soon as possible (particularly in perlite and clay pebbles). After transplanting, timed irrigation cycles might be spaced a little closer together.

Environmental conditions for plant growth

The 70 range is the ideal daily temperature for tomatoes and other warm-season crops. When temperatures rise over mid-80, growth will slow down. High temperatures and/or temperature fluctuations can also lead to fruit quality problems like blossom end rot. These problems are more prevalent when there is an imbalance between water loss through transpiration to the atmosphere and water circulation bringing calcium through the plant and fruit. Fruiting crops need intense light all the time to perform well and lower disease risks. Shading may be required on the hottest and brightest summer days to minimize heat and prevent heat buildup. Full sun is required during other times of the year. High yields may be restricted by low winter light levels, which are frequently insufficient for optimum growth and productivity. Small-scale growers could decide to put up with slower growth and lower crop yields in exchange for fresh vegetables during the off-season or to work with school schedules. Controlling the nutrient solution or the environment can frequently boost growth, productivity, and quality. Sunlight, however, cannot be replaced by any amount of solution management or plant training.

Plant Training

Plant support and training

Supporting fruit-bearing plants is crucial for their well-being and output. Therefore, you'll need overhead wires or bars in your growing area so that plants may be trained and controlled vertically. Using plastic thread, tomatoes, cucumbers, peppers, and eggplants will all be trained vertically (other natural string fibers may break down during the crop cycle). However, utilizing a spool or bobbin that may release more string, as detailed below, can be simpler to manage and maintain over a longer length of time. Plastic twine for tomatoes can be connected to the top horizontal wire. The main stem is fastened to the vertical string with plastic clips.

Figure 57: Bobbin used to hold extra twine and secure a vining crop to the overhead support wire.

Source: CropKing Inc.

The majority of indeterminate tomatoes are trained to one main stem, which usually has three leaves in between each flower/fruit cluster. Lateral branches develop at each of the leaf axils, however, they need to be cut off to keep the plant as a single stem. For the duration of the plant, these suckers (lateral branches) are cut off. Keeping plants to a single main stem keeps them manageable in size and helps the plant balance the amount of energy used for fruit vs leaf and stem growth. To avoid having to remove huge suckers, this procedure should be carried out frequently (about weekly). The plant may suffer harm, suffer significant wounds, and lose energy if they are removed.

Figure 58: A tomato plant showing the plastic twine and clip to secure the vine.

Source: CropKing Inc.

In order to keep the plants more manageable, cucumbers are also trained vertically to twine and the majority of side branches are also cut off. The development of tendrils is one of the difficulties in controlling cucumbers. These are frequently removed in order to stop them from growing on other plants, support wires, and posing management issues. Each leaf axil (the site where leaves connect to the main stem) produces one or more flowers that will eventually develop into fruits. Because of their rapid growth, cucumbers are frequently left to continue bearing fruit until the main stem reaches the top support wire. At that point, the stem loops over the support and is then allowed to continue growing back down to the ground, extending the time between harvests. With the lateral branches cut, peppers and eggplants are also taught to twine strung from top wires. However, two leaders are frequently permitted to grow on pepper and eggplant plants. Peppers and eggplants typically develop two branches in a "V" at each branch. Both branches are allowed to grow at the plant's base and the first V, and the strongest or most desirable stem is chosen to preserve two growing sites. This means that in hydroponic systems, twice as many bobbins of twine will be required for peppers and eggplants than there are for indeterminate tomatoes.

Plant management practices

Two key management techniques are utilized frequently in soilless tomato growing systems. The first is referred to as lowering and leaning. The tomato plants will soon get to the top horizontal wire due to the main stem's steady growth. The growing point can be reduced while still maintaining growth and harvest. This entails moving (leaning) the growing point down the wire while more string is gradually released from the bobbin or spool. The extra thread is then let go, gently lowering the plant. The plant may be lowered without damaging the stem thanks to the movement along the top horizontal wire. The frequency of leaning and lowering varies based on the crop and the environment, but it typically occurs once each week. Cherry, grape, and roma tomatoes frequently need to be leaned and lowered since they are generally leggier. Older leaves are removed to maintain hygienic standards and balance the plant's energy requirements (often right before leaning and lowering). It's crucial to understand how many to remove and when to do so. A plant typically has 10–14 leaves at a time, and the number of leaves can serve as a general indicator. The sort of tomato can affect this. Consider the location of the ripening cluster of fruit on the plant as another helpful statistic. Typically, leaves below the most mature cluster are removed since, after the fruit starts to ripen, they are of little nutritional use to the developing fruit. Another clue is where the bottom leaves are placed as a result of leaning and lowering. Leaves on the floor should be cleared because they are unwelcome. Leaves can be prevented from contacting the floor with the help of a top horizontal wire that is high enough. Because it takes time and cucumbers grow swiftly, leaning and lowering are less frequent. These plants are trained vertically until they reach the support wire because the stems of pepper and eggplant are too brittle for such techniques.

Figure 59: Twine support for young crops

Source: agriplasticscommunity.com/

Pollination

While seedless cucumbers don't need pollination, tomatoes, eggplants, and peppers need the process. Bumblebees are commonly employed for pollination in commercial buildings, but smaller greenhouses typically avoid using them since the smallest bumblebee hives include a colony that is too huge. As a result of their frequent visits, a large population of bees will harm flowers. When there are few plants, hives can be expensive, making manual pollination the most practical and economical method. You can buy commercial pollinators or even use an electric toothbrush as a pollinator. The best conditions for pollination—when the relative humidity is between 60 and 80 percent—are three to four times per week. As a result, pollination typically takes place around the middle of the day. When fruiting crops are grown outside, the wind aids in pollination and pollinators have access to the plants.

Harvest and Storage

To have the best flavor and appearance, harvest fruit when it is fully ripe. Tomatoes are physiologically fully ripe when the color has reached 20% or higher. But most farmers only pick when the redness is between 60% and 80%. Cucumbers can be picked at almost any stage the farmer chooses. Fruit should not be left on the plant for an excessive amount of time since it can lower the energy plants can devote to younger, developing fruit. When the peppers reach the desired hue, they are picked. To have a superior texture, eggplants are frequently removed before the seeds form. Harvesting eggplant while the skin still has a robust, shiny appearance is a vital indicator. Below 50 F, tomatoes, eggplants, and cucumbers are all chilling sensitive. Fruit can be refrigerated between 55- and 60-degrees Fahrenheit to delay ripening and prevent chilling harm. Warmer conditions will lengthen respiration and shorten shelf life. If necessary, peppers can be kept a little cooler (45 F) to increase shelf life. It is advised to store all of these fruits at high relative humidity levels (>95 percent) to minimize water loss. The seedless cucumber with thin skin is a good example of this. Cucumbers frequently need to be individually wrapped quickly after being picked since water loss is so quick and harmful. Smaller seedless mini cucumbers are less likely to require wrapping since their skins do not lose water as quickly.

GROWING HYDROPONIC STRAWBERRIES

One of the most consumed fruits worldwide is strawberries. A ripe red strawberry is so loving because of the perfect harmony of sweet and tart flavor. Americans consume 4 pounds of strawberries annually, according to sources. Hydroponics is not the easiest method for growing fruits. Strawberries, on the other hand, are unique in this situation. Strawberries can be cultivated year-round in great quantities using the hydroponic technique. A high-value crop known for its flavor, color, and nutritional value is the strawberry. But cultivating strawberries can be difficult. The essentials that should be taken into account when producing strawberries in hydroponic production systems are provided in the following information.

Variety

Farmers like hydroponically cultivating strawberries from runners. Strawberry seedlings can take a very long time to grow. Up to 3 years are needed for a strawberry seed to become a plant. where cultivation requires intensive attention and harvesting. However, the fruits produced by these plants are not of very high quality. Planting from seed is not an option for someone who wishes to grow strawberries in a year. He must run through the runner to farm. The most common and widely used method, particularly in hydroponics, is runner propagation.

Strawberries come in a variety of varieties, and you can start them in a number of various methods. However, some are superior to others for home gardeners. What matters most is how quickly you want the strawberry plants to produce fruit. Some can start bearing fruit right away, while others can take up to three years. We'll stick to strawberry kinds with a shorter period to harvest as it's quite unlikely you'll want to wait that long to pick. An ever-bearing species or day-neutral strain is what you want if you're seeking a type of strawberries that consistently produces fruit. While day-neutral strains can continue to bear fruit throughout their lives, ever-bearing strains can readily generate up to three harvests per year. Here are some of your finest options if you're looking to cultivate a certain variety of strawberry hydroponically:

- Douglas
- Tioga
- Tuft
- Chandler
- Brighton
- Red Gauntlet

Figure 60: Hydroponic strawberry.

Source: hortamericas.com/blog/

Growing System

Finding a suitable hydroponic system is the first step. Price, plant count, and performance of these all vary substantially. Therefore, you must exercise utmost vigilance while buying your system. Stick with an ebb-and-flow, deep water culture (DWC), or hydroponic drip system when initially starting out. You can create your own system, but there are many fantastic ones that are already built and ready to use. Budget growers can get started quickly and for a small portion of the cost of a conventional system by using a garden tray, reservoir, water pump, and a few other auxiliary hydro components. The actual construction of the system is rather simple. The only thing left to do is to set up your water reservoir beneath the tray, which is where your plants will develop. Then, configure your pump and timer to transfer water from the reservoir into your grow tray, feeding and watering your berries as needed. You should also think about the growing media you'll employ. Growstones, clay pebbles, coconut coir, or rockwool are a few popular options. There are other further expanding media formats available. Additionally, you'll likely need some hydroponic nutrients to feed your plants properly and encourage their maximum growth. But as the grower, this is entirely up to you. While there are alternative hydroponic techniques for growing strawberries, the wick method is the simplest of them all. Strawberry plants are a good example of a little plant for the Wick method.

Media

There are many different hydroponic growing mediums available, and they all essentially do the same thing. However, we advise keeping with something simple and accessible, like coco coir. Your strawberries' pH or nutritional value won't be affected by coco, which is absolutely inert. It is affordable, widely accessible, and incredibly beginner-friendly. To improve aeration and water drainage, you might wish to add some perlite. Aeration of the strawberry root zone is important. As a result, it is advised that strawberries be grown on soil that can offer the root zone at least 15% porosity. Based on a short study conducted in 2008 by the University of Arizona, which looked at various substrates and substrate combinations:

- granulated Rockwool
- 100% coconut coir
- 50% coconut coir + 50% perlite
- 70% coco nut coir + 30% perlite

When coconut coir made up the whole growing media, strawberry plants' growth and yield were constrained. Their theory about the significance of oxygen availability was supported by the fact that better results were obtained when Rockwool was utilized, and the best results were obtained when perlite was used. The University of Arizona currently mixes 50% perlite, 25% coconut coir, and 25% peat as its growing medium in its greenhouse. Since their supply water has a high pH, the addition of peat will allow for better pH control.

Substrate Management

Hydroponically grown strawberries require some sort of fertilizer solution to keep them alive, whether you are growing organically or not. Together with secondary and micronutrients, your strawberries require a lot of nitrogen, potassium, and phosphorus. Nutrients in liquid form must be used. It's simpler for newbies to begin with synthetics and advance from there because organics have a tendency to clump and clog hydroponic systems. General Hydroponics Nutrients is one of our favorite product lines. You may easily follow the feeding schedules provided by these features so that you always know exactly how much and when to feed your strawberry plants. They are affordable, of the highest quality, and suitable for beginners. You may decide whether you want organic or chemical fertilizer and what form you want it in when choosing the right hydroponic fertilizers for strawberries. We advise utilizing a liquid fertilizer that's simple to add into your reservoir to make things simple for newcomers. In this manner, you may more easily manage the fertilizer requirements of your plants. Make sure you use chemical fertilizer that is especially made for hydroponic use when utilizing fertilizers. This is due to the fact that plants cultivated without soil don't often respond well to general-purpose chemical fertilizers like those you may get at a hardware store. When it comes to organic hydroponic nutrients for strawberries, things become a little more difficult. This is due to the fact that it can be difficult to obtain a source of organic nutrients that can be dissolved in water. As soon as you locate a source, you must confirm that the nutrient levels are sufficient to support plant development. The following are the two major groups of essential nutrients:

- **Macronutrients:** Nitrogen, Phosphorus, and Potassium
- **Trace elements (Micronutrients):** those act as secondary nutrients – Calcium, Magnesium, and Sulphur

It is up to you to decide which system you want to use to advance. However, it's crucial to discover strawberry nutrients that are suitable for the environment in which they will be produced. It's also important to note that you must monitor the pH of your reservoir. You should make sure that the pH of your nutrient solution is between since strawberries love a little acidic environment. There are numerous strawberry hydroponic feeding formulas available. However, Farm Fox and General Hydroponics remain the most well-liked brands. They will supply the essential macronutrients, including potassium, phosphorus, nitrogen, and trace elements. The kind of setup the strawberries are in will determine how long you need to feed them. It's best to avoid feeding your plants too much in a hydroponic garden. Salts may accumulate excessively as a result, converting into hazardous chemicals and ultimately killing the organisms. Generally speaking, feeding your strawberry plants twice a week is advised. Every two to three weeks, or every five to six days if there are numerous plants, you should change the nutrition solution. If you apply too much fertilizer at once, your soil will become depleted and won't support new growth. To prevent this, sprinkle a small amount of water mixed with the diluted liquid fertilizer to the soil outside margins. From there, take it all in (this will ensure that nutrients are going down as far as they need). Finally, follow any further instructions provided by your maker!

The ideal pH range for hydroponic systems is 5.5 to 6.5, which is perfect for strawberry plants. Use a simple pH meter and aim for a pH of between 5.8 and 6.2 while cultivating strawberries for the best results. Using test strips is the most straightforward technique to determine pH because they are relatively cheap and have a high success rate. Instead of just "high/low," you may purchase a pH meter that is more accurate and provides values on a real scale that ranges from one to fourteen.

Management Practices

Being a fruit, a hydroponically grown strawberry nevertheless needs some attention. There are a few different ways to do this, but the most frequent ones include pollination, pruning, and fruit rot.

Pollination: Bees and other insects can pollinate strawberry plants. Since nature operates in a similar manner, you should aim to follow suit for the greatest outcomes. Cross-pollination is encouraged by pollination, which boosts plant fruit output. Hydroponically grown strawberries can be pollinated by moving pollen from the male plant's stamen, where it is visible, to the female sections of another strawberry using a small paintbrush or cotton swab.

Pruning: Plants require room to expand, therefore pruning is necessary. Don't worry if your berry bush has developed too many flimsy branches; just follow these easy steps: Cut a couple of stems through at the base (do not clip them). Per clipped stem, you should snip off roughly 25% of the leaves. Even if trimming doesn't considerably affect yield, it will keep your entire garden organized. Stale berries, it can assist in preventing the growth of fungus, which lowers quality.

Fruit Rot: Any rotting on the outside or inside of your strawberry is referred to as fruit rot. Overwatering and having too much fertilizer close to the roots are the primary culprits for this. Your strawberries will taste bitter after eating them as a result. Fortunately, harvesting your strawberries as soon as they ripen will help you avoid these fruit rot diseases. Neem oil applications can aid in the eradication of fruit rot diseases should they develop.

Watering: Make sure your nutrition solution has been properly diluted and that you aren't watering your plants too frequently or seldom. The fruit of the winter is strawberries. They enjoy a temperature range of 15 to 27 degrees Celsius during their growing season since they do best in chilly environments. The agricultural cultivation in Arizona observed that the strawberry fruit output had increased during the winter nights when the temperature was between 10 and 20 degrees Celsius.

Disease Control: One of the various ways a plant might contract a disease is by being too near other plants that also have the ailment. Give your strawberry plants plenty of area to grow so that they don't spread out more. Additionally, avoid overwatering and overfertilizing them to prevent your strawberries from tasting bitter when you eat them. Additionally, always use potting soil or mix of high grades.

Humidity Control: For strawberries to grow, the ideal humidity level must be maintained. Calcium intake is impacted by low humidity, which has an impact on photosynthesis and fruit quality. A minimum of 60 to 75 percent relative air humidity must be maintained; a higher nocturnal humidity encourages nutrient flow during the day.

Lighting: Strawberries grown hydroponically require at least 8 to 12 hours of direct sunlight daily. If you reside somewhere that receives little sunlight throughout the winter. Purchasing artificial lighting such as LED or HID lights is recommended.

Harvesting: At least three-quarters of the fruit's surface must have turned from green to red for the berries to be ready for picking. When lightly squeezed, the berries will likewise feel supple and give slightly. To avoid harming the plant, you can collect them by hand or with a pair of tiny scissors. They can be consumed immediately, used in a dish, or kept in the refrigerator for up to a week.

When eating fruits or vegetables cultivated hydroponically, many people detect changes in sweetness, texture, and taste. This has to do with the nutrient solution, which varies considerably amongst hydroponic systems and gives rise to various tastes. Another factor is that strawberries have a lower surface area than plants grown in soil, which means that less airflow (oxygen) passes through them. That raises their sweetness as well.

Steps To Plan A Large-Scale Hydroponics Farm

The future of agriculture is hydroponics. One of the most important advances in horticulture since the invention of irrigation is hydroponics, which can be used to grow food for human consumption or medicinal herbs. More than 80% of the world's population is predicted to reside in urban regions by the year 2050. This will make it extremely difficult for us to feed everyone using current means. Additionally, freshwater resources are at an all-time low and will continue to decline in the upcoming years, leaving less water available for the cultivation of these commodities. For each of these significant issues, a solution is provided by hydroponic farms. When compared to conventional farming, hydroponics saves up to 99 percent of the water required for cultivation and yields 13% more per square meter. Large-scale hydroponic operations now have a great deal of demand in the urban environment. However, how can this occur? What procedures must be planned in order to put a large-scale hydroponics farm together?

Figure 61: Large-scale hydroponic farm.

Source: linkedin.com/pulse/

Step 1: Location: You need to give the location of your hydroponics business careful thought. Do you want to sell your fruit at nearby farmers' markets or do you want to sell it to nearby restaurants? Finding a place farther from the city with a lower lease cost will be simple if selling at farmers' markets. It would be preferable for you to select a place closer to the locations you will be focusing on supplying if you are especially supplying eateries.

Step 2: Crop Choice: Before beginning any large-scale hydroponic farm, you must carefully determine what crop you will plant. This will depend on a number of factors, including local supply and demand, production costs, and crop profit margins. Although spinach and lettuce are among the easiest crops to raise and have some of the best profit margins, your location may already be oversaturated with these products depending on how many other business owners came up with the idea before you. Prior to beginning your hydroponic farm, you will need to conduct a significant amount of market research because this is what will keep you in business. Contact the Department of Agriculture, Environment, and Rural Affairs in your area. It's crucial to decide if you'll concentrate your efforts on a single crop or aim for a variety of crops to offer various markets. Both have been done and can be done successfully, but most experts agree that it is far better to concentrate on one particular crop and learn how to maximize its production.

Step 3: System Choice: The system you choose will directly depend on the space you have available and the crop you are growing. Crops that grow shorter and faster, like lettuce and spinach, thrive in large flood tables or aeroponic systems. These plants develop quickly and stay short, allowing the planter to cram a lot of individual plants—hundreds of them—into a compact space while yet getting plenty of light. Larger plants, such as tomatoes or fruits, require more space to expand both horizontally and vertically. NFT systems or deep-water culture DWC will be beneficial for these plants. Additionally, the setup fee must be taken into account. The cost of constructing huge hydroponic farms will be substantially different from small size operations. Unlike small operations, which are often built of pre-molded plastic, flood tables can be made from metal and wood.

Step 4: Choosing Lighting: Large hydroponic farms should take lighting into account because the power consumption margins would significantly affect the cost of these systems. While costing almost three times as much as HID bulbs, LED lights have a longer lifespan and use far less electricity per lumen. Consider a 600-watt LED light that costs roughly 60 USD less per month than a 1000-watt HPS bulb while producing an equivalent amount of light (depending on local energy costs). The lifetime savings of an LED (over a 25-year period) is greater than (16,400 USD) per light when compared to the annual replacement expenses of HPS and MH lighting at (85 USD per bulb). Long-term investment here is enormous given that a large-scale hydroponics operation will require several lights. Planning should take into account whether you have the resources to choose LEDs right away or whether you will need to save development costs and pay more in the long run.

Figure 62: Artificial LED Lighting for a large-scale hydroponic farm.

Source: linkedin.com/pulse/

REFERENCES

1. Abad, M., Fornes, F., Carrión, C., Noguera, V., Noguera, P., Maquieira, Á., & Puchades, R. (2005). Physical properties of various coconut coir dusts compared to peat. HortScience, 40(7), 2138-2144.
2. Acuña, T. B., & Wade, L. J. (2012). Genotype× environment interactions for root depth of wheat. Field Crops Research, 137, 117-125.
3. Alexander, P. D., Bragg, N. C., & Stuart, J. H. C. (2015, September). Informed decision making: a tool to aid the decision-making process of sourcing responsible growing media throughout the UK supply chain. In International Symposium on Growing Media, Composting and Substrate Analysis-SusGro2015 1168 (pp. 93-100).
4. Allen, R. G., Pereira, L. S., Raes, D., & Smith, M. (1998). Crop evapotranspiration-Guidelines for computing crop water requirements-FAO Irrigation and drainage paper 56. Fao, Rome, 300(9), D05109.
5. Anon, 2011. Sri Lanka: Coconut Production and Export. ,http://www.cri.gov.lk/web/images/stories/statistics/satistics_on_coconut_oil_palm.pdf..
6. Arya, L. M., & Paris, J. F. (1981). A physicoempirical model to predict the soil moisture characteristic from particle-size distribution and bulk density data. Soil Science Society of America Journal, 45(6), 1023-1030.
7. Azad, K. N., Salam, M. A., & Azad, K. N. (2016). Aquaponics in Bangladesh: current status and future prospects. Journal of Bioscience and Agriculture Research, 7(02), 669-677.

8. Bao, Y., Aggarwal, P., Robbins, N. E., Sturrock, C. J., Thompson, M. C., Tan, H. Q., ... & Dinneny, J. R. (2014). Plant roots use a patterning mechanism to position lateral root branches toward available water. Proceedings of the National Academy of Sciences, 111(25), 9319-9324.
9. Bar-Tal, A., Saha, U. K., Raviv, M., & Tuller, M. (2019). Inorganic and synthetic organic components of soilless culture and potting mixtures. In Soilless culture (pp. 259-301). Elsevier.
10. Benoit, F., & Ceustermans, N. (1993, October). A decade of research on ecologically sound substrates. In International Seminar on Soilless Culture Technology Protected Crops Mild Winter Climates 408 (pp. 17-30).
11. Birch, P. (1979). Concrete costs repaid by easier lettuce management. . Grower, May 3, 31–35.
12. Blok, C., Van der Salm, C., Hofland-Zijlstra, J., Streminska, M., Eveleens, B., Regelink, I., ... & Visser, R. (2017). Biochar for horticultural rooting media improvement: evaluation of biochar from gasification and slow pyrolysis. Agronomy, 7(1), 6.
13. Boodley, J. W. (1973). Boron deficiency and petal necrosis of Indianapolis white chrysanthemum.
14. Boyer, C. R., Gallagher, T. V., Gilliam, C. H., Fain, G. B., Torbert, H. A., & Sibley, J. L. (2012). Description of clean chip residual forest harvest and its availability for horticultural uses in the southeastern United States. HortTechnology, 22(3), 381-387.
15. Bragg, N. C. (1997, April). The commercial development of a sustainable peat alternative substrate from locally derived industrial by-products. In International Symposium on Composting & Use of Composted Material in Horticulture 469 (pp. 61-70).
16. Bunt, A. C. (1990, September). The relationship of oxygen diffusion rate to the air-filled porosity of potting substrates. In II Symposium on Horticultural Substrates and their Analysis, XXIII IHC 294 (pp. 215-224).
17. Bures, S., & Pokorny, F. A. (1991). Equation for estimating shrinkage in binary mixtures of container media. HortScience, 26(8), 1087-1087.
18. Burés, S., Gago, M. C., Morales, O., Marfà, O., & Martínez, F. X. (1997a). Water characterization in granular materials. Acta Horticulturae, 389-396.

19. Burés, S., Pokorny, F. A., Landau, D. P., & Ferrenberg, A. M. (1993). Computer simulation of volume shrinkage after mixing container media components. Journal of the American Society for Horticultural Science, 118(6), 757-761.
20. Bussell, W. T., & Mckennie, S. (2004). Rockwool in horticulture, and its importance and sustainable use in New Zealand. New Zealand journal of crop and horticultural science, 32(1), 29-37.
21. Bustamante, M. A., Restrepo, A. P., Alburquerque, J. A., Pérez-Murcia, M. D., Paredes, C., Moral, R., & Bernal, M. P. (2013). Recycling of anaerobic digestates by composting: effect of the bulking agent used. Journal of Cleaner Production, 47, 61-69.
22. Carlile, W. R. (2003, February). Changes in organic growing media during storage. In South Pacific Soilless Culture Conference-SPSCC 648 (pp. 153-159).
23. Carmassi, G., Incrocci, L., Maggini, R., Malorgio, F., Tognoni, F., & Pardossi, A. (2007). An aggregated model for water requirements of greenhouse tomato grown in closed rockwool culture with saline water. Agricultural water management, 88(1-3), 73-82.
24. Cary, J. W., & Hayden, C. W. (1973). An index for soil pore size distribution. Geoderma, 9(4), 249-256.
25. Castano, R., Borrero, C., Trillas, M. I., & Avilés, M. (2013). Selection of biological control agents against tomato Fusarium wilt and evaluation in greenhouse conditions of two selected agents in three growing media. BioControl, 58(1), 105-116.
26. Chamindu Deepagoda, T. K. K., Moldrup, P., Tuller, M., Pedersen, M., Chen Lopez, J. C., Wollesen de Jonge, L., ... & Komatsu, T. (2013a). Gas diffusivity-based design and characterization of greenhouse growth substrates. Vadose Zone Journal, 12(3), 1-13.
27. Chen Lopez, J. C. (2011). Characterization, Simulation, Analysis and Management of Hydraulic Properties of Greenhouse Plant Growth Substrates.
28. Cole, J. C., & Dunn, D. E. (2002). Expanded polystyrene as a substitute for perlite in rooting substrate. Journal of Environmental Horticulture, 20(1), 7-10.
29. Cooper, A. J. (1975). Crop production in recirculating nutrient solution. Scientia Horticulturae, 3(3), 251-258.

30. Crippa, L., Zaccheo, P., & Orfeo, D. (2013, October). Utilization of the solid fraction of digestate from anaerobic digestion as container media substrate. In International Symposium on Growing Media, Composting and Substrate Analysis 1013 (pp. 367-373).
31. Da Silva, F. F. (1991). Static and dynamic characterization of container media for irrigation management. Hebrew University of Jerusalem, Faculty of Agriculture.
32. Da Silva, F. F., Wallach, R., & Chen, Y. (1994, September). Hydraulic properties of rockwool slabs used as substrates in horticulture. In International Symposium on Growing Media & Plant Nutrition in Horticulture 401 (pp. 71-76).
33. Dale, J. (2018). The Evolution of the Sediment Regime in a Large Open Coast Managed Realignment Site: A Case Study of the Medmerry Managed Realignment Site, UK (Doctoral dissertation, University of Brighton).
34. De Graaf, R. (1988). Automation of the water supply of glass house crops by means of calculating the transpiration and measuring the amount of drainage Water. Acta Hort, 229, 219-231.
35. Deepagoda, T. C., Lopez, J. C. C., Møldrup, P., de Jonge, L. W., & Tuller, M. (2013). Integral parameters for characterizing water, energy, and aeration properties of soilless plant growth media. Journal of Hydrology, 502, 120-127.
36. Dickinson, K., & Carlile, W. R. (1994, September). The storage properties of wood-based peat-free growing media. In International Symposium on Growing Media & Plant Nutrition in Horticulture 401 (pp. 89-96).
37. Doğan, M., & Alkan, M. (2004). Some physicochemical properties of perlite as an adsorbent.
38. Domeño, I., Irigoyen, I., & Muro, J. (2010). New wood fibre substrates characterization and evaluation in hydroponic tomato culture. European Journal of Horticultural Science, 75(2), 89.
39. Durand, M., Porcheron, B., Hennion, N., Maurousset, L., Lemoine, R., & Pourtau, N. (2016). Water deficit enhances C export to the roots in *Arabidopsis thaliana* plants with contribution of sucrose transporters in both shoot and roots. Plant Physiology, 170(3), 1460-1479.

40. Durner, W. (1994). Hydraulic conductivity estimation for soils with heterogeneous pore structure. Water resources research, 30(2), 211-223.
41. Epstein, E. (1972). Mineral nutrition of plants: principles and perspectives.
42. Evans, M. R., Konduru, S., & Stamps, R. H. (1996). Source variation in physical and chemical properties of coconut coir dust. HortScience, 31(6), 965-967.
43. Evans, M. R., Konduru, S., & Stamps, R. H. (1996). Source variation in physical and chemical properties of coconut coir dust. HortScience, 31(6), 965-967.
44. Fayolle, L., Noble, R., Coventry, E., Aime, S., & Alabouvette, C. (2006). Eradication of Plasmodiophora brassicae during composting of wastes. Plant Pathology, 55(4), 553-558.
45. FLJAW, V. (1977). Growing horticultural crops in rockwool and nutrient film. [Conference paper]. In Fourth International Congress on Soilless Culture. Las Palmas (Spain). 25 Oct 1976.
46. Fornes, F., Belda, R. M., Abad, M., Noguera, P., Puchades, R., Maquieira, A., & Noguera, V. (2003). The microstructure of coconut coir dusts for use as alternatives to peat in soilless growing media. Australian Journal of Experimental Agriculture, 43(9), 1171-1179.
47. Gaag, D. J. V. D., & Wever, G. (2005). Conduciveness of different soilless growing media to Pythium root and crown rot of cucumber under near-commercial conditions. European journal of plant pathology, 112(1), 31-41.
48. Gericke, W. F. (1937). Hydroponics—Crop production in liquid culture media. Science, 85(2198), 177-178.
49. Giuffrida, F., & Consoli, S. (2016). Reusing perlite substrates in soilless cultivation: Analysis of particle size, hydraulic properties, and solarization effects. Journal of Irrigation and Drainage Engineering, 142(2), 04015047.
50. Giuffrida, F., Leonardi, C., & Marfà, O. (2007, October). Substrate reuse in tomato soilless cultivation. In International Symposium on High Technology for Greenhouse System Management: Greensys2007 801 (pp. 1577-1582).

51. Gruda, N., Tucher, S. V., & Schnitzler, W. H. (2000). N-immobilization by wood fibre substrates in the production of tomato transplants (Lycopersicon lycopersicum (L.) Karst. ex Farw.). Angewandte Botanik, 74(1/2), 32-37.
52. Grunert, O., Hernandez-Sanabria, E., Vilchez-Vargas, R., Jauregui, R., Pieper, D. H., Perneel, M., ... & Boon, N. (2016). Mineral and organic growing media have distinct community structure, stability and functionality in soilless culture systems. Scientific reports, 6(1), 1-14.
53. Han, J. J., Lee, S. B., Park, Y. G., & Jeong, B. R. (2014). Flower yield and quality of two rose cultivars grown in phenolic foam LC slab and phenolic foam RC slab in comparison to perlite and rockwool slab. Horticulture, Environment, and Biotechnology, 55(2), 70-78.
54. Handreck, K. A., Black, N. D., & Black, N. (2002). Growing media for ornamental plants and turf. UNSW press.
55. Hanna, H. Y. (2005). Properly recycled perlite saves money, does not reduce greenhouse tomato yield, and can be reused for many years. HortTechnology, 15(2), 342-345.
56. Hanna, H. Y. (2010). Reducing time and expense to recycle perlite for repeat use in greenhouse tomato operations. Horttechnology, 20(4), 746-750.
57. Hardgrave, M. (1994, September). An evaluation of polyurethane foam as a reusable substrate for hydroponic cucumber production. In International Symposium on Growing Media & Plant Nutrition in Horticulture 401 (pp. 201-208).
58. Harland, J., Lane, S., & Price, D. (1997, May). Further experiences with recycled zeolite as a substrate for the sweet pepper crop. In International Symposium on Growing Media and Hydroponics 481 (pp. 187-196).
59. Harrelson, T., Warren, S. L., & Bilderback, T. E. (2004). How do you manage aged versus fresh pine bark. In Proc. Southern Nursery Assn. Annu. Conf. 49th Annu. Rpt (pp. 63-65).
60. Heller, H., Bar-Tal, A., Assouline, S., Narkis, K., Suryano, S., de la Forge, A., ... & Tsohar, D. (2015). The effects of container geometry on water and heat regimes in soilless culture: Lettuce as a case study. Irrigation Science, 33(1), 53-65.

61. Hinsinger, P., Plassard, C., Tang, C., & Jaillard, B. (2003). Origins of root-mediated pH changes in the rhizosphere and their responses to environmental constraints: a review. Plant and soil, 248(1), 43-59.
62. Hose, E., Clarkson, D. T., Steudle, E., Schreiber, L., & Hartung, W. (2001). The exodermis: a variable apoplastic barrier. Journal of Experimental Botany, 52(365), 2245-2264.
63. İncemehmetoğlu, A. (2012). Investigation the effects of different support medium on product with nutrient film technique (Master›s thesis, Middle East Technical University).
64. Ingram, D. L., Ruter, J. M., & Martin, C. A. (2015). Characterization and impact of supraoptimal root-zone temperatures in container-grown plants. HortScience, 50(4), 530-539.
65. Johansson, J. (1978). Effects of nutrient levels on growth, flowering and leaf nutrient content of greenhouse roses. Acta Agriculturae Scandinavica, 28(4), 363-386.
66. Jones, P., & Martin, M. (2003). A review of the literature on the occurrence and survival of pathogens of animals and humans in green compost. WRAP standards report. The Waste and Resources Action Programme, Oxon, United Kingdom.
67. Jones, S. B., & Or, D. (1998). Design of porous media for optimal gas and liquid fluxes to plant roots. Soil Science Society of America Journal, 62(3), 563-573.
68. Jury, W. A., & Flühler, H. (1992). Transport of chemicals through soil: Mechanisms, models, and field applications. Advances in Agronomy, 47, 141-201.
69. Kaufhold, S., Reese, A., Schwiebacher, W., Dohrmann, R., Grathoff, G. H., Warr, L. N., ... & Ufer, K. (2014). Porosity and distribution of water in perlite from the island of Milos, Greece. SpringerPlus, 3(1), 1-10.
70. Kivinen, E., & Pakarinen, P. (1980, August). Peatland areas and the proportion of virgin peatlands in different countries. In Proceedings of the 6th International Peat Congress, Duluth (pp. 52-54).
71. Koltermann, C. E., & Gorelick, S. M. (1995). Fractional packing model for hydraulic conductivity derived from sediment mixtures. Water Resources Research, 31(12), 3283-3297.
72. Kratky, B. A. (2005). Growing lettuce in non-aerated, non-circulated hydroponic systems. Journal of vegetable science, 11(2), 35-42.

73. Kudo, Y., Kotani, M., Tomita, M., & Aizawa, Y. (2009). Effects of rock wool on the lungs evaluated by magnetometry and biopersistence test. Journal of Occupational Medicine and Toxicology, 4(1), 1-7.
74. Lambers, H., Cramer, M. D., Shane, M. W., Wouterlood, M., Poot, P., & Veneklaas, E. J. (2003). Introduction: structure and functioning of cluster roots and plant responses to phosphate deficiency. Plant and Soil, 248(1/2), ix-xix.
75. Lamont, J. R., & Elliott, G. C. (2016). Anaerobically digested dairy fiber in soilless potting media for poinsettias. International Journal of Recycling of Organic Waste in Agriculture, 5(2), 173-177.
76. Lawrence, W. J. C., & Newell, J. (1939). Seed and potting composts. Seed and potting composts.
77. Leoni, S., Grudina, R., & Pisanu, B. (1994). A new cultivation system of the greenhouse tomato. High density aeroponics. Informatore Agrario (Italy).
78. Luxmoore, R. J. (1981). Micro-, meso-, and macroporosity of soil. Soil Sci. Soc. Am. J.;(United States), 45(3).
79. MacConnell, C. B., & Collins, H. P. (2009). Utilization of re-processed anaerobically digested fiber from dairy manure as a container media substrate. In International Symposium on Growing Media 2007 819 (pp. 279-286).
80. Malorgio, F., Scacco, M., Tognoni, F., & Pardossi, A. (2001). Effect of nutrient concentration and water regime on cut rose production grown in hydroponic system. Acta horticulturae. 0.17660/ ActaHortic.2001.559.46.
81. Marfa, O., Martínez, A., Orozco, R., Serrano, L., & Martínez, F. X. (1992, September). The Use Of Fine-grade Perlites In Lettuce Bag Cultures. Ii. Physical Properties, Rheologic Effects And Productivity. In International Symposium on Horticultural Substrates other than Soil in situ 342 (pp. 339-348).
82. Martin, C. A., & Ingram, D. L. (1991). Evaluation of thermal properties and effect of irrigation on temperature dynamics in container media. Journal of Environmental Horticulture, 9(1), 24-28.
83. Meinken, E. (1996, September). Accumulation of nutrients in expanded clay used for indoor plantings. In International Symposium Growing Media and Plant Nutrition in Horticulture 450 (pp. 321-328).

84. Mengel, K., & Kirkby, E. A. (2012). Principles of plant nutrition. Springer Science & Business Media.

85. Moncada, A., Caracciolo, G., & D'Anna, F. (2007, October). Effects of substrate solarization on tomato soilless cultivation. In International Symposium on High Technology for Greenhouse System Management: Greensys2007 801 (pp. 1485-1492).

86. Montero, J. I., Antón, A., Munoz, P., & Lorenzo, P. (2001). Transpiration from geranium grown under high temperatures and low humidities in greenhouses. Agricultural and Forest Meteorology, 107(4), 323-332.

87. Morgan, J. V. (1985). Ireland: current research and development. . In Hydroponic Worldwide: State of the AU15 Art in Soilless Crop Production (A.J. Savage ed.). pp. 116–122.

88. Nash, M. A., & Pokorny, F. A. (1992). Prediction of water-retention of milled pine bark-sand potting media from laboratory analyses of individual components. Communications in soil science and plant analysis, 23(9-10), 929-937.

89. Natalie B. & Robert H., (2019). W 844-C-Tomatoes, Peppers and Cucumbers in Small-Scale Soilless and Hydroponics Systems

90. Nelson, P. V., Oh, Y. M., & Cassel, D. K. (2001, September). Changes in physical properties of coir dust substrates during crop production. In International Symposium on Growing Media and Hydroponics 644 (pp. 261-268).

91. Noble, R., & Roberts, S. J. (2004). Eradication of plant pathogens and nematodes during composting: a review. Plant pathology, 53(5), 548-568.

92. Noble, R., & Roberts, S. J. (2004). Eradication of plant pathogens and nematodes during composting: a review. Plant pathology, 53(5), 548-568.

93. Olympios, C. M. (1992, March). Soilless media under protected cultivation rockwool, peat, perlite and other substrates. In Symposium on Soil and Soilless Media under Protected Cultivation in Mild Winter Climates 323 (pp. 215-234).

94. Pardossi, A., Carmassi, G., Diara, C., Incrocci, L., Maggini, R., & Massa, D. (2011). Fertigation and substrate management in closed soilless culture. Pisa: University of Pisa.

95. Penningsfeld, F. (1966). und Kurzmann, P. Hydrokultur und Torfkultur. Verlag Eugen Ulmer, Stuttgart, 1966.

96. Pérez-Hernández, A., Porcel-Rodríguez, E., & Gómez-Vázquez, J. (2017). Survival of Fusarium solani f. sp. cucurbitae and fungicide application, soil solarization, and biosolarization for control of crown and foot rot of zucchini squash. Plant Disease, 101(8), 1507-1514.
97. Pickering, H. W., Menzies, N. W., & Hunter, M. N. (2002). Zeolite/rock phosphate—a novel slow-release phosphorus fertiliser for potted plant production. Scientia Horticulturae, 94(3-4), 333-343.
98. Pokorny, F. A., & Henny, B. K. (1984). Construction of a milled pine bark and sand potting medium from component particles. II: Medium synthesis. Journal of the American Society for Horticultural Science, 109(6), 774-776.
99. Pokorny, F. A., & Wetzstein, H. Y. (1984). Internal porosity, water availability, and root penetration of pine bark particles. HortScience, 19(3), 447-449.
100. Pokorny, F. A., Gibson, P. G., & Dunavent, M. G. (1986). Prediction of bulk density of pine bark and/or sand potting media from laboratory analyses of individual components. Journal of the American Society for Horticultural Science. 111(1):8—11.
101. Ponchia, G., Passoni, M., Bonato, S., Nicoletto, C., Sambo, P., & Zanin, G. (2015, September). Evaluation of compost and anaerobic digestion residues as a component of growing media for ornamental shrub production. In International Symposium on Growing Media, Composting and Substrate Analysis-SusGro2015 1168 (pp. 71-78).
102. Pond, W. G., & Mumpton, F. A. (1984). Zeo-agriculture: Use of natural zeolites in agriculture and aquaculture. Westview press.
103. Prasad, M. (1996, September). Nitrogen fixation of various material from a number of European countries by three nitrogen fixation tests. In International Symposium Growing Media and Plant Nutrition in Horticulture 450 (pp. 353-362).
104. Prasad, M. (1996, September). Physical, chemical and biological properties of coir dust. In International Symposium Growing Media and Plant Nutrition in Horticulture 450 (pp. 21-30).
105. Prasad, M., & Maher, M. J. (2003, February). Stability of peat alternatives and use of moderately decomposed peat as a structure builder in growing media. In South Pacific Soilless Culture Conference-SPSCC 648 (pp. 145-151).

106. Prasad, M., & O'Shea, J. (1997, May). Relative breakdown of peat and non-peat growing media. In International Symposium on Growing Media and Hydroponics 481 (pp. 121-128).

107. Prasad, M., Ní Chualáin, D., & Hynes, C. (2015, September). The effect of moisture content on fraction recovery and the development of a rapid field test for moisture determination of sod peat (sod moss and bin sods). In International Symposium on Growing Media, Composting and Substrate Analysis-SusGro2015 1168 (pp. 387-394).

108. Prasad, M., Tzortzakis, N., & McDaniel, N. (2018). Chemical characterization of biochar and assessment of the nutrient dynamics by means of preliminary plant growth tests. Journal of environmental management, 216, 89-95.

109. Pronk, A. A. (1994, September). Composted vegetable, fruit and garden waste as a substitute for peat in container-grown nursery stock. In International Symposium on Growing Media & Plant Nutrition in Horticulture 401 (pp. 473-480).

110. Raviv, M., & Blom, T. J. (2001). The effect of water availability and quality on photosynthesis and productivity of soilless-grown cut roses. Scientia Horticulturae, 88(4), 257-276.

111. Raviv, M., Lieth, H., & Bar-Tal, A. (Eds.). (2019). Soilless culture: Theory and practice: Theory and practice. Elsevier.

112. Raviv, M., Silber, A., & Medina, S. (1998). The effect of medium disinfestation on cut rose productivity and on some chemical properties of tuff. Scientia horticulturae, 74(4), 285-293.

113. Ridout, M. E., & Tripepi, R. R. (2009, June). Initial chemical and physical properties of potting mixes amended with anaerobically digested cattle biosolids. In International Symposium on Growing Media and Composting 891 (pp. 167-172).

114. Robinson, D. W., Lamb, J. G. D., & Horticultural Education Association. (1975). Peat in horticulture.

115. Romero-Gámez, M., Antón, A., Leyva, R., & Suárez-Rey, E. M. (2017). Inclusion of uncertainty in the LCA comparison of different cherry tomato production scenarios. The International Journal of Life Cycle Assessment, 22(5), 798-811.

116. Ross, P. R., Thenmozhi, P., & Raj, S. P. (2010). Utilization of coir waste as a soilless medium. Advances in Biological Research, 4(4), 198-200.

117. Rupp, L. A., & Dudley, L. M. (1989). Iron availability in rockwool may affect rose nutrition. HortScience, 24(2), 258-260.
118. Sahin, U., Ercisli, S., Anapali, O., & Esitken, A. (2003, February). Regional distribution and some physico-chemical and physical properties of some substrates used in horticulture in Turkey. In South Pacific Soilless Culture Conference-SPSCC 648 (pp. 177-183).
119. Schindler, U., Lischeid, G., & Müller, L. (2016). Hydraulic performance of horticultural substrates—3. Impact of substrate composition and ingredients. Horticulturae, 3(1), 7.
120. Schmilewski, G. (2015, September). Growing media constituents used in the EU in 2013. In International Symposium on Growing Media, Composting and Substrate Analysis-SusGro2015 1168 (pp. 85-92).
121. Schröder, F. G., & Förster, R. (2000, May). Polyester fleece as substrate for soilless culture. In World Congress on Soilless Culture: Agriculture in the Coming Millennium 554 (pp. 67-74).
122. Sherman, J. D. (1999). Synthetic zeolites and other microporous oxide molecular sieves. Proceedings of the National Academy of Sciences, 96(7), 3471-3478.
123. Silber, A. (1991). Chemical properties and surface reactions of pyroclastic materials from Mt. Peres, the Golan Heights (Doctoral dissertation, Ph. D thesis, Hebrew Univ. of Jerusalem, Israel).
124. Silber, A., & Raviv, M. (1996). Effects on chemical surface properties of tuff by growing rose plants. Plant and soil, 186(2), 353-360.
125. Singh, H., & Dunn, B. (2016). Electrical conductivity and pH guide for hydroponics. Oklahoma Cooperative Extension Service.
126. Smith, D. L. (1992). Rockwool in horticulture. Grower Books, 1987: ISBN: 0901361828, 9780901361820.
127. Smith, J. (2005). Hydroponics: A Practical guide for the soilless grower. HortTechnology, 15(3), 731-731.
128. Soffer, H., & Levinger, D. (1980). The Ein Gedi system-research and development of a hydroponic system. In Proceedings, Fifth International Congress on Soilless Culture. Secretariat, ISOSC.
129. Sonneveld, C., & Voogt, W. (1985). Studies on the application of iron to some glasshouse vegetables grown in soilless culture. Plant and soil, 85(1), 55-64.

130. Sonneveld, C., Voogt, W., & Spaans, L. (1997, May). A universal algorithm for calculation of nutrient solutions. In International Symposium on Growing Media and Hydroponics 481 (pp. 331-340).

131. Spomer, L. A. (1974). Optimizing container soil amendment: The" threshold proportion" and prediction of porosity. HortSci, 9, 532-533.

132. Steele, T. W., Lanser, J. A. N. I. C. E., & Sangster, N. O. R. M. A. (1990). Isolation of Legionella longbeachae serogroup 1 from potting mixes. Applied and environmental microbiology, 56(1), 49-53.

133. Van Iersel, M. (1997). Root restriction effects on growth and development of salvia (Salvia splendens). HortScience, 32(7), 1186-1190.

134. Van Iersel, M. W., & Dove, S. K. (2013, June). Temporal dynamics of oxygen concentrations in a peat-perlite substrate. In International Symposium on Growing Media and Soilless Cultivation 1034 (pp. 355-361).

135. Van Os, E. A., & Stanghellini, C. (2001). Water reuse in greenhouse horticulture. In Proceedings EURO-Summer school, Program' Water and resource Recovery in Industry', Wageningen (NL), 26-31 August, 8 pp.

136. Van Os, E. A., Gieling, T. H., & Lieth, J. H. (2019). Technical equipment in soilless production systems. In Soilless culture (pp. 587-635). Elsevier.

137. Verwer, F. L. J. A. (1977, September). Research and results with horticultural crops grown in rockwool and nutrient film. In Symposium on Production of Protected Crops in Peat and other Media 82 (pp. 141-148).

138. von Post, L. (1922). Sveriges geologiska undersöknings torvinventering och några av dess hittills vunna resultat.

139. Wald, C. (2017). The new economy of excrement. Nature, 549(7671), 146-148.

140. Walter, A., Spies, H., Terjung, S., Küsters, R., Kirchgessner, N., & Schurr, U. (2002). Spatio-temporal dynamics of expansion growth in roots: automatic quantification of diurnal course and temperature response by digital image sequence processing. Journal of experimental botany, 53(369), 689-698.

141. Wever, G., & Van Leeuwen, A. A. (1994, September). Measuring mechanical properties of growing media and the influence of cucumber cultivation on these properties. In International Symposium on Growing Media & Plant Nutrition in Horticulture 401 (pp. 27-34).
142. Wik, R. M., Fisher, P. R., Kopsell, D. A., & Argo, W. R. (2006). Iron form and concentration affect nutrition of container-grown Pelargonium and Calibrachoa. HortScience, 41(1), 244-251.
143. Yeager, T., Gilliam, C., Bilderback, T. E., Fare, D., Niemiera, A., & Tilt, K. (1997). Best management practices: Guide for producing container-grown plants. Southern Nursery Assoc., Marietta, GA.
144. Zaccheo, P., Crippa, L., & Cattivello, C. (2013, June). Liming power of different particle fractions of biochar. In International Symposium on Growing Media and Soilless Cultivation 1034 (pp. 363-368).
145. Zhang, H., & Forde, B. G. (2000). Regulation of Arabidopsis root development by nitrate availability. Journal of experimental botany, 51-59.

INDEX